THE GOLDILOCKS PLANET

Jan Zalasiewicz is Senior Lecturer in Geology at the University of Leicester, and was formerly with the British Geological Society. A field geologist, palaeontologist, and stratigrapher, he teaches various aspects of geology and Earth history to undergraduate and post-graduate students, and is a researcher into fossil eco-systems and environments across half a billion years of geological time. He is the author of *The Earth After Us* (OUP, 2008) and *The Planet in a Pebble* (OUP, 2010).

Mark Williams is Professor of Palaeobiology at the University of Leicester, and was formerly a scientist with the British Antarctic Survey. He teaches many aspects of geology but particularly climate change over geological timescales. His research interests lie in how the fossil record shows changes in the Earth's climate. He has published over a hundred papers in scientific journals.

JAN ZALASIEWICZ
& MARK WILLIAMS

OXFORD
UNIVERSITY PRESS

Great Clarendon Street, Oxford, OX2 6DP

Oxford University Press is a department of the University of Oxford.
It furthers the University's objective of excellence in research, scholarship,
and education by publishing worldwide. Oxford is a registered trade mark of
Oxford University Press in the UK and in certain other countries

First Edition published in 2012
First published in paperback 2013

Impression: 1

Published in the United States of America by Oxford University Press
198 Madison Avenue, New York, NY 10016, United States of America

British Library Cataloguing in Publication Data
Data available

ISBN 978–0–19–959357–6 (hbk)
ISBN 978–0–19–968350–5 (pbk)

Printed in Great Britain on acid-free paper by
Clays Ltd, St Ives plc

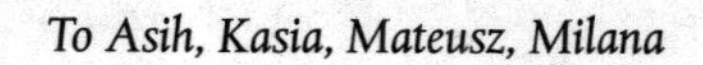
To Asih, Kasia, Mateusz, Milana

Contents

Acknowledgements

This book grew out of a long-held interest we both have in trying to work out what the world used to be like in olden times—in the days of our human ancestors, and in those nigh-countless days before humans evolved, when the world was very different. What did much to determine the nature of those prehistoric worlds, we have found, was climate. Not so much in whether any particular primordial landscape was delicately sun-kissed, or grey and misty, or windy and rain-swept. That is part of it, of course, but the effect of climate is far more fundamental. It governs the balance between land and sea, the shape of the landscape, the pattern of sea currents —and the nature of living communities. Hence this exercise in telling the story of Earth's climate—from its beginnings to its changeable present and uncertain future.

This book has been supported by those at Oxford University Press, who have patiently and amiably converted a rather untidy assortment of words and pictures into a proper book, with Latha Menon's perceptive editing in particular steering the narrative into shape. Our thanks, too, to colleagues who have read all or part of the narrative, and/or who have provided information or illustrations: Harry Dowsett, John Smellie, Steve Hesselbo, Euan Nisbet, Ian

Fairchild, Eric Wolff, Ellen Bazeley-White, Danielle Stoll, Jerry Davies, Howard Armstrong, Appy Sluis, and Joe Hiess.

Our own explorations of the Earth's past climate and environment have been influenced by colleagues past and present, who are almost too numerous to name. Among them, though, colleagues at the University of Leicester, including Mike Branney, Tiffany Barry, Andy Saunders, Dick Aldridge, David Siveter, Sarah Gabbott, Dickson Cunningham, Mike Norry, John Hudson, Alex Page, Peter Long and Melanie Leng; and at other institutions, including Adrian Rushton, Alan Haywood, Dan Hill, Andy Johnson, Ulrich Salzmann, Beth Okamura, Barrie Rickards, Thijs Vandenbroucke, Phil Gibbard, Nick Shackleton, Andrew Gale, Stewart Molyneux, Colin Waters, Felix Gradstein, Dan Condon, Jacques Grinevald, Paul Crutzen, Will Steffen, Mike Ellis, Sir Crispin Tickell, and Jean Vannier. We must add, though, that errors, omissions, oversights and excessive oversimplifications in this narrative are down to us alone.

Finally, we'd like to thank those family and friends who in years distant and near have variously supported and encouraged us in our exploration of the Earth's past: our parents—Feliks, Irena, Les and Doreen, early teachers and mentors (right back to encouragement of explorations into dinosaurs, trilobites and such a very long while ago). Creating the climate to nurture curiosity is a valuable thing.

List of Illustrations

Prologue

What might we see in walking across the Earth in Cretaceous times? One would admire the magnificent reptiles, of course—this was the Age of Dinosaurs—but also remark on how different that world was. The tropics would be hot—uncomfortably hot and humid for our taste. Travelling north, to the polar regions, we would see little trace of ice. Instead, there would be dense conifer forests of cypress and redwood, with gingkos and ferns, adapted to live in perpetual day for half a year and perpetual night for half a year. For sightseeing, there would be no need for high-specification polar gear: a good pullover would do nicely (don't forget the umbrella, though, as there would be a *lot* of rain).

Or, imagine tracking backwards three hundred million years to the Carboniferous, with tropical jungles at the equator (where Yorkshire would have been, and Pennsylvania too), but with a gigantic mass of ice on South America and southern Africa—then at the South Pole. Or we might go further back to the Silurian, when the oceans were often so still that oxygen did not penetrate their depths, and life on the sea floor perished. Or yet further back to what is now called the Cryogenian, seven hundred million years ago, when ice spread to cover the world, then melted abruptly amid such climatic turbulence that hurricanes raged across the Earth. These are vastly different states of our planet that we would be visiting—so much so, that they almost represent different planets.

We are used to our weather changing from day to day, within the general limits of our local climate. We know, too, that Earth's climate today varies from place to place, from the burning heat of the deserts of the Sahara and the Gobi, to the brief summers and bitter winters of Siberia and Arctic Canada, to the perpetual icehouse of the centre of Antarctica, to the mild and rain-swept climate of an English winter.

But what was Earth's climate like, long ago, before humans appeared on this planet? It might seem a frivolous question. But today we are increasingly aware that global climate—and therefore our familiar weather patterns—are changing, as our human civilization alters the landscape, the Earth's freight of animals and plants, the chemistry of both atmosphere and oceans, and even the whiteness of polar ice. How far will these changes go, and how fast might they be?

When the future is uncertain, then the past may provide some indication of what may happen in the future. This is why history has always been such a good preparation for politicians who have to negotiate the vagaries of human behaviour in peace and in war—and why such negotiations can be fruitless when history is ignored or forgotten.

The history of climate extends far beyond human history, and stretches back some four and a half billion years to our planet's formation, at about the time our Sun's nuclear fires ignited. That history can be read from clues preserved in the Earth's strata. The evidence is abundant, though always incomplete—and often baffling, tantalizing, seemingly contradictory. Geologists, though, are becoming ever more ingenious at interrogating this evidence, and the story of the Earth's climate is now being reconstructed in ever greater detail and—one hopes—in an ever more faithful representation of what the Earth was actually like, millions (and even billions) of years ago. Given something as complex as climate, these reconstructions—and the search for mechanisms as to *how* and *why* climate changed—involve not only geologists but chemists, biologists, physicists, oceanographers, atmospheric scientists, glaciologists, computer modellers, and many others. Almost every scientific discipline (and a few that lie outside the sciences, in the humanities) is represented in this quest.

The accumulated evidence, and the story, is now enormous—of far greater scale and scope than can be represented in this modest volume. Let us say that this book has about 90,000 words—including conjunctions, quantifiers, indefinite and definite articles and adjectives. With a near five-billion-year history, that comes out to one word for every fifty thousand years. This is an absurdly compressed synthesis. Nevertheless, there are patterns to discern. As regards climate, history has repeated itself on this planet, sometimes many times over—perhaps never in exactly the same way, but similarly enough to begin to draw some reasonable conclusions as to how and even as to why climate changed as it did.

There is also, unsurprisingly, more evidence and better evidence the nearer to our own times we get. Something approaching half the book is devoted, therefore, to the last five million years—that is one-tenth of one per cent—of Earth history, and the many climate switchbacks of the current Ice Age and in the times that immediately preceded it. But more ancient times can still provide wonderful evidence of how the Earth's climate, and hence its surface environment, can be utterly transformed. The rapid growth of an enormous ice sheet on Africa at the end of the Ordovician Period, almost half a billion years ago, and its collapse a geological instant later, is a prime example. Or there are examples when climate seems to have stayed remarkably stable over times that really do pass human understanding, such as the largely uneventful billion-year span (as regards climate, and much else) in the middle of the Proterozoic Era. The times of (apparent) stability have as much to teach us as the times when global climate patterns, repeatedly, seem to turn on a sixpence (or on a dime, if you prefer).

The sheer variety of evidence out there telling us about climates past is in itself remarkable: from the shells of minute protozoans, to the courses of the planets, to beautifully fossilized leaves, to invisible atomic patterns within strata—even to the assiduous toilet habits of certain desert-dwelling rodents. So too are the efforts that scientists have made to collect and analyse that data, across all parts of the world, from the highest mountains to below the depths of the ocean floor. This has involved ingenuity, luck (both good and bad), the taking up of undignified positions (quite common, this, in our experience), serendipity, cussedness, danger, triumph—and, now and again, tragedy. Through all of this, the story has built up.

We hope some flavour of these aspects of the story of Earth's climate comes through in these pages. It is a story that is incomplete as well as abbreviated. It might, though, provide a backcloth to discussions of contemporary climate change—and it is an infinitely absorbing topic in itself.

The approach we take in these pages is broadly chronological. This is the natural approach to a geologist. Blessed as we are with so much time, we need to treat it seriously and systematically. A brief word on the time spans we shall meet is therefore in order, before we visit the beginning of the world.

A Brief Word on Time

In this book we have to navigate through a stretch of time that amounts to a little over four and a half billion years. That is the whole of the age of the Earth (and, incidentally, about a third of the age of the Universe). So as not to become hopelessly lost within such a quite inhuman time span, we need to follow the geologists' approach to locating events within it, of changes in landscapes, oceans, and biology—all the things that have influenced climate on our planet.

Geologists do not simply use years as a scale of reference, even in multiples of millions and billions. The reason for this is partly historical. The early geologists had gone a long way to developing a fully functional timescale for Earth history—without having any good idea of how long their time units were. This is because they had not yet discovered the means to date events in the past by exploiting the natural radioactivity of minerals. Their time units were essentially based on the recognition of successive dynasties of animals and plants, preserved as fossils within strata.

Geologists still use this timescale (albeit an updated and refined version), partly because the numerical timescale is still not everywhere precisely defined, and partly because the units—Jurassic, Carboniferous, Pleistocene, and so on—are familiar and form a more effective common language than do numbers.

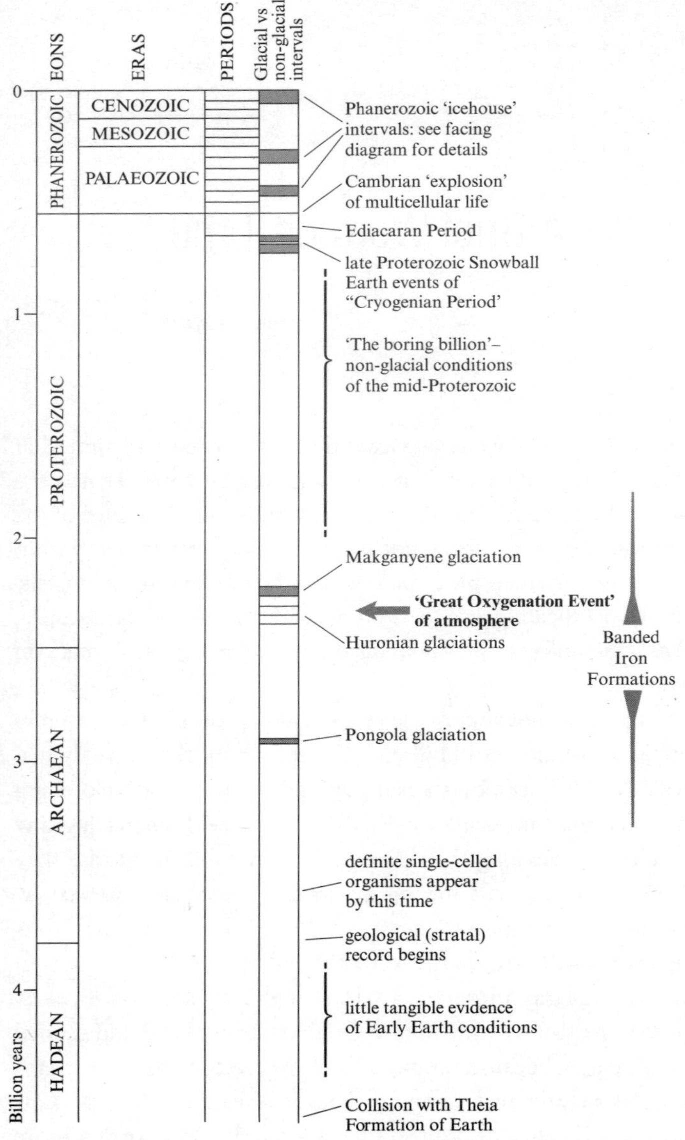

FIG. 1. The Geological Timescale and some of the key events in the history of the Earth's climate.

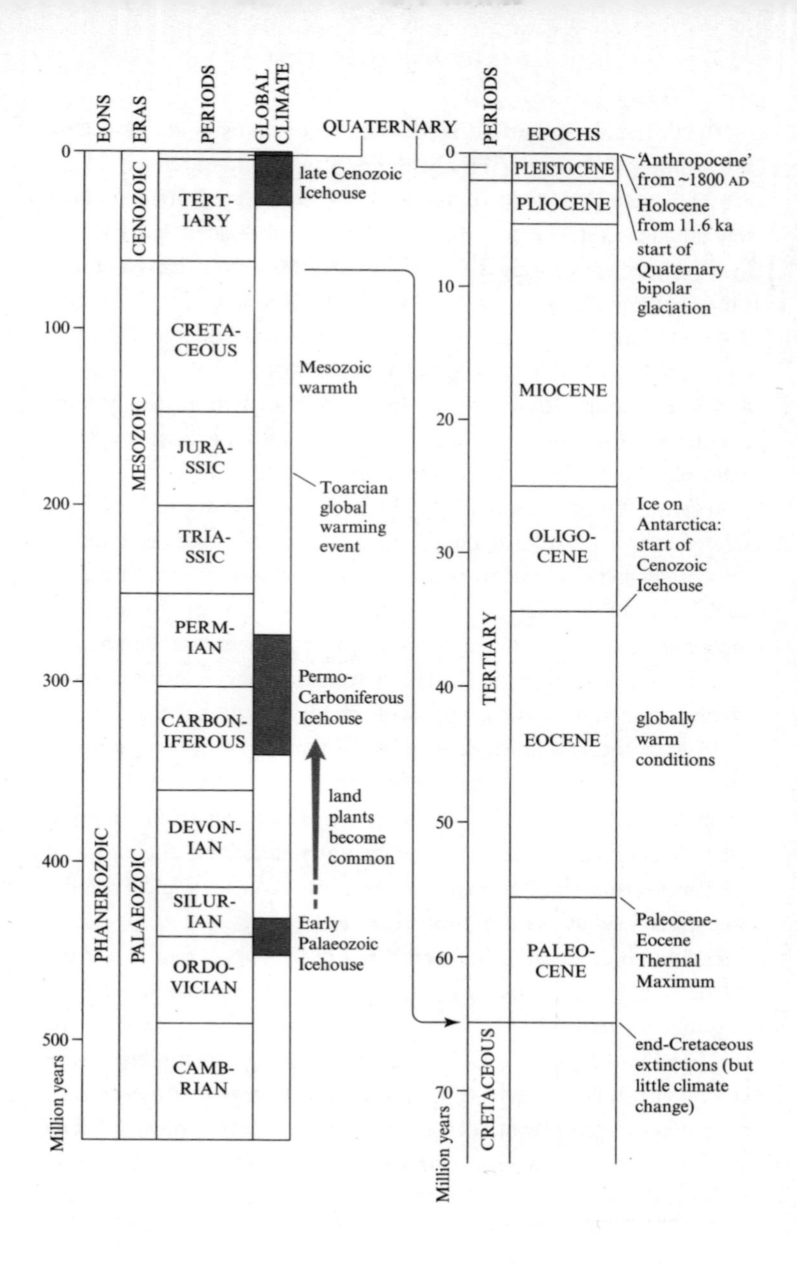

EONS
ERAS
PERIODS
GLOBAL CLIMATE
QUATERNARY
PERIODS
EPOCHS
0
100
200
300
400
500
Million years
PHANEROZOIC
CENOZOIC
MESOZOIC
PALAEOZOIC
TERT-IARY
CRETA-CEOUS
JURA-SSIC
TRIA-SSIC
PERM-IAN
CARBON-IFEROUS
DEVON-IAN
SILUR-IAN
ORDO-VICIAN
CAMB-RIAN
late Cenozoic Icehouse
Mesozoic warmth
Toarcian global warming event
Permo-Carboniferous Icehouse
land plants become common
Early Palaeozoic Icehouse
0
10
20
30
40
50
60
70
Million years
TERTIARY
CRETACEOUS
PLEISTOCENE
PLIOCENE
MIOCENE
OLIGO-CENE
EOCENE
PALEO-CENE
'Anthropocene' from ~1800 AD
Holocene from 11.6 ka
start of Quaternary bipolar glaciation
Ice on Antarctica: start of Cenozoic Icehouse
globally warm conditions
Paleocene-Eocene Thermal Maximum
end-Cretaceous extinctions (but little climate change)

The timescale, in outline, is shown in Fig. 1. In using it, there are a few things to remember. Firstly, the timescale is hierarchical. There are a few very large units of time—eons—that are subdivided into a few smaller units—eras—that are in turn subdivided successively into periods, epochs, and then ages. So, the Mesozoic Era is a unit of time lasting almost 200 million years that is a part of the 542-million-year-long Phanerozoic Eon, within which we still live. The Mesozoic Era is divided into the familiar periods of the Triassic, Jurassic, and Cretaceous. These are in turn subdivided more finely, into epochs and then into ages; we shall refer to a few of those units (for example, the Toarcian Age of the Jurassic) within this book.[1]

Most of the time units in the Phanerozoic Eon are defined by a reference level in strata at a particular, carefully selected point that is informally called a 'golden spike'. This reference level usually marks some distinctive change, mostly in the fossils but increasingly by recognizing chemical or magnetic changes in the strata which indicate significant changes in the Earth System. Many of those changes involved substantial and long-lasting changes in global climate.

Prior to the Phanerozoic—and so for most of Earth history—there are few fossils and so it is harder to locate suitable reference levels for 'golden spikes'. For this more ancient Earth history, the time divisions are (for the time being) defined on the basis of numbers—thus the boundary between the Archaean and Proterozoic Eons is set at 2,500 million years ago.

Finally, there is one time term used in this book that is not yet formally part of the Geological Timescale (though conceivably it might be formalized in the years to come). This is the Anthropocene, that headlines Chapter 10. It is increasingly being used to refer to the last two centuries, in which human actions have come to dominate the surface of this planet.[2] Part of the evidence for the Anthropocene involves human effects on climate—so the narrative of this book may help you decide your own perspective on this concept.

Primordial Climate

Alternative histories

We are lucky, on Earth. We are lucky because we—as complex and self-aware organisms—are here. We are sustained, given air to breathe, and water, and food, by a very ancient planet: a planet past its midpoint, a planet that is nearer death than birth.

Our species is a latecomer. It took some three billion years to bridge the gap from a single-celled organism (originating in this planet's youth) to a multicellular one, and then a little over half a billion more to arrive at the diversity of species on Earth today, including *Homo sapiens*. In all this time, the chain of life has remained unbroken. The Earth has been consistently habitable, with an atmosphere, and land, and oceans. Since life began, our planet has never been truly deep-frozen, nor have the oceans boiled away.

The Earth is the Goldilocks planet. One recalls, here, the children's story, where the young heroine of that name walks into the house of the three bears, and in their absence tries out successively their bowls of porridge, their chairs, and their beds. Each time the first and second choices are too hot or cold, large or small, hard or soft—and the third choice is *just right*. The Earth has been, so far and all in all, *just right* for life: not just right at any one time, but continuously so for three billion years. There have, though, been some close calls: times of mass extinction. But, life has always clung on to bloom once more. That makes the Earth's history more remarkable than any children's story.

Other planets have not been so lucky. Mars seems to have been a planet with an appreciable atmosphere, and—at least intermittently—running water over its surface,[1] and may even have begun to incubate life. But the atmosphere was stripped away by the solar wind. Its early lakes and rivers became acid, charged with sulphates. Then, most of the water evaporated and was carried off into space; what little was left became locked away as permafrost and in thin ice-caps. Mars does have weather, including spectacular, planet-wide dust-storms. It might even harbour a few simple microbes. But it will never be a green and pleasant land.

Venus, too, probably had a temperate climate once. Indeed, in its early history it probably started with about as much water as the Earth possesses—only to lose both that water, and any chance of maintaining a climate fit for life.[2] At some point in its history, the temperature rose high enough for water vapour to begin to leak into its upper atmosphere, where the solar wind could carry it away. There remain chemical fingerprints of that loss of Venusian water. The minute traces of water left in Venus's atmosphere are now hugely enriched in 'heavy water'—that is, in water possessing the heavy isotope of hydrogen (deuterium, that has a nucleus with both a neutron and a proton, as opposed to the single proton of normal hydrogen). It is the type of water molecule that is easier to retain within the atmosphere, while almost all the normal, lighter water molecules were whisked off into outer space.

As the water disappeared, carbon dioxide could no longer be removed from the atmosphere by forming acid rain and reacting with rocks. The carbon dioxide in the Venusian atmosphere built up to enormous levels. Virtually by itself, it makes up an atmosphere ninety times denser than Earth's (and one that is laced, for good measure, with sulphuric acid, droplets of which produce the thick, light-reflecting Venusian clouds). The heat-trapping properties of this gas converted the planet's surface into a furnace at over 400 °C—and this even though Venus receives less sunlight at the surface than does the Earth, because of its reflective clouds. So, there is a climate, and weather on Venus—but not one that we would wish to experience on Earth.

The climate of Hades

When did the Earth's climate begin? This was probably not when the Earth first formed, 4,567 million years (give or take a million years or so) after it had rapidly accreted, in just a few million years, from some of the rocky debris whirling around the just-formed Solar System. Whatever envelope of liquids and gases it had around itself, and whatever climate was emerging, was rendered irrelevant by an event that changed everything.

The high-speed collision of the Earth with a Mars-sized planet, christened Theia, perhaps as much as 100 million years into the Earth's history, is inferred to be the most likely means by which we acquired our Moon, which coalesced from the spray of molten and gaseous silicates flung out by the impact.[3] The Moon, with no water and no atmosphere, has always been devoid of climate and of weather.

The Earth, immediately after collision, would have been a magma ocean, with a turbulent atmosphere of silicate vapour. Then, over the subsequent thousands and millions of years, it would have transformed itself into the mysterious world of the Hadean, an eon that lasted almost a billion years. It is a long history with which we have almost no contact, for almost no physical traces remain of that early

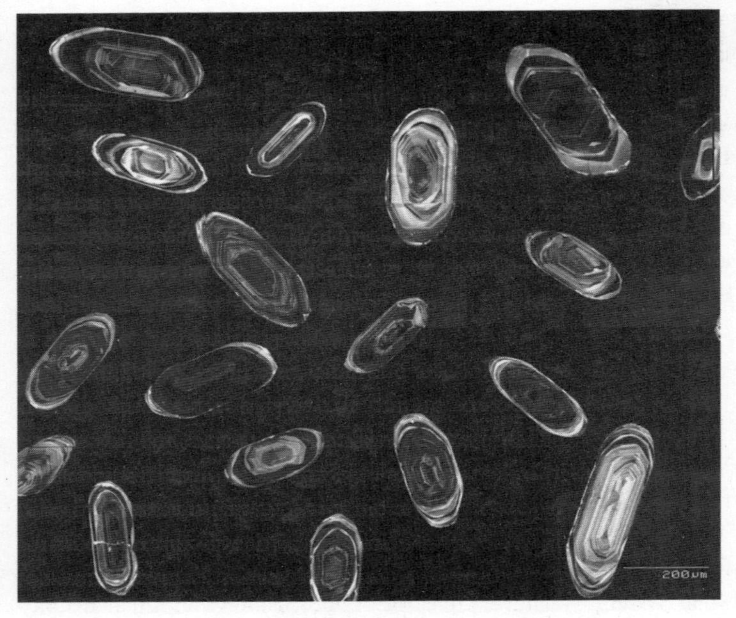

FIG. 2. Ancient zircons from 3.88-billion-year-old metamorphic rocks of the Iltsaq Gneiss complex, West Greenland. The scale bar represents one-fifth of a millimetre.

world: virtually all the surface crust of that time has been thoroughly reworked and transformed into younger rocks.

Most speculation upon the state of the world then is based on a scattering of tiny crystals, the world's oldest, that exceed four billion years in age.[4] They come from the Jack Hills, out in the near-desert of Western Australia, too arid to support anything other than skinks and geckos and a few feral goats. The rocks there represent sands and pebble layers that accumulated on some primeval beach or river floodplain, some three billion years ago. Within them, though, are exceedingly rare crystals of zircon, each just a fraction of a millimetre across, which had been washed in by those ancient rivers from yet older rock outcrops, now vanished. They are almost all that we have from that early Earth.

The chemistry and structure of these zircon crystals show that they formed about 7 kilometres underground, at temperatures of somewhere between 700 and 800 °C. And, because they are our best link to that early time, they have been interrogated with all the ingenuity that geologists can muster, to tell us something of what the Hadean world was like, high above at the surface. It was one in which the peculiarly terrestrial engine of plate tectonics may not yet have started. The chemistry of those zircons suggests, rather, that the magma ocean cooled to form a kind of 'proto-crust', a planet-wide shell made of the rock types basalt and anorthosite (the latter a distinctive pale rock, formed almost entirely of one of the forms of the mineral feldspar), in something like the way that the Moon did. Volcanic edifices built up on this crust and, deep within these, the Jack Hills zircons were formed.

That primeval world was one that, even then, may have had liquid water at its surface—and therefore, presumably, weather systems that included rain. The evidence for this lies in the pattern of the different types of oxygen atoms—isotopes—within the mineral structure of the zircon. The particular pattern involves the ratio of the 'normal' light isotope, of atomic weight 16 (^{16}O, having a total of eight protons and eight neutrons in its nucleus) and a heavier isotope, ^{18}O (with eight protons and ten neutrons). These isotopes tend to be separated (or fractionated, as it is termed) at the Earth's surface by processes such as evaporation: a water molecule bearing the heavy oxygen atom will tend to evaporate less easily than one with a light oxygen atom.

The Hadean Jack Hills zircons are enriched in heavy oxygen, and from this it has been suggested that these zircons crystallized from magmas that had some interchange with bodies of liquid water, high above on the surface, that had been undergoing fractionation. This is slender evidence on which to infer the world's first climate. There may, perhaps, have been other processes deep underground that sorted those oxygen isotopes into the pattern they eventually took, within those tiny zircon crystals. The suggestion of oceans remains a tantalizing hint. But it is all that we have, for now.

During the Hadean, the framework of the Earth's climate was defined. The zircons hint that the Earth's hydrosphere was already in place. The original building blocks of the Earth probably included masses of ice, which simply dissolved into the silicate material of the Earth. Some of this original water undoubtedly still circulates at deep levels far beneath the Earth's surface, and is the source of some of the steam that rises from volcanic eruptions. More water would have arrived as comets and icy meteorites crashed into the Earth. This was a violent and eventful time, as the solar system was being cleaned of much of its debris by the simple expedient of smashing it into planets. The larger impacts on the Earth, of meteorites a few hundred kilometres in diameter, would have imparted enough energy to vaporize all or part of the early oceans.

Streaming out from the Sun is not only the radiation that brings light and warmth, but also a 'wind' of charged particles that can strip away an atmosphere. It is likely that part of the Earth's envelope of water and gases was removed by the solar wind early in the history of the planet. At some early point, too, a system of electric 'eddy' currents arose in the liquid metal at the Earth's core, generating a protective magnetic field around the whole planet.[5] From that time until today, the Earth's magnetic field has acted as a shield, which deflects the abrasive solar wind.

The impact with Theia also set up the basic machinery that regulates our climate. It defined the spin and the tilt of the Earth, and set up the whirling Earth–Moon system (the energy of which has been dissipated, exceedingly slowly, by the tides, the Moon edging gradually away from the Earth). This, in turn, controls the day–night cycle and the seasons, resulting from the way in which heat and light are distributed around the Earth. Some time in the Hadean, too, some form of plate tectonics probably started up: parts of that early proto-crust were stretched apart by vigorous convection loops in the partially molten mantle beneath (a mantle that was then some 200 °C hotter than today), and other slabs of crust slid down into that mantle, setting up a planetary machine that still functions today.[6] That machine has created the basins in which the ocean waters accumulate, producing great stores of heat carried by currents that are in turn disrupted and redirected as they flow around the continents.

Together, these elements have resulted in a dynamic climate machine, and direct evidence of its action may be seen among some of the world's oldest strata.

The evidence of the earliest strata

The earliest fossilized landscapes on Earth may be found in the hearts of the most ancient continents, in Greenland, Canada, and Australia. They are the remains of shallow sea floors, lakes, and rivers, between 3 and 3.8 billion years old. These show no sign that the world then possessed extensive ice sheets. Signs of a thoroughgoing glaciation are generally clear. Where moving streams of ice grind across a landscape they leave characteristic masses of debris and land surfaces with the tell-tale signs of ice-scratching and gouging. For nearly a billion years of Archaean time, these signs are absent from the strata that remain. This is not to say that ice was completely absent: there may have been mountain glaciers, traces of which have completely vanished because mountains are areas of erosion. Moreover, the Archaean record is patchy and poor: little is left, and the strata that have been preserved have mostly been crumpled, heated, and pervasively altered by mountain-building movements. Nevertheless, had there been extensive, long-lasting glaciations (as occurred later) we should have found evidence of them by now.

Evidence of liquid water at the surface abounds, though, in the form of ripples sculpted by waves and dunes swept by water currents, and as curious convex build-ups of laminae of calcium carbonate, a

FIG. 3. Vestiges of early microbial life: Precambrian stromatolites from the Bonahaven Formation, Islay, Scotland.

metre or so in diameter, within the limestones. These are called stromatolites, and the oldest examples are some 3.5 billion years old. Rarely, very similar structures can be found growing today in some shallow seas: Shark Bay in Australia, for instance, currently hosts a fine assemblage of these limestone masses. Stromatolites form by microbial mats trapping fine limy sediment, and so the ancient examples are generally taken as evidence that there was both water and primitive life even then.[7] The modern examples typically grow in conditions too hostile for most multicellular creatures. In Shark Bay, for instance, they grow in a pool of water that has been isolated from the sea for some three thousand years, and in which the water has become highly salty in the hot climate—so salty that creatures such as snails and crustaceans cannot live there. In normal seas today such creatures simply eat through any young and hopeful stromatolite.

The early Earth of the Archaean, then, was neither a completely frozen one, nor one so hot that all the planet's water turned to steam and all life-forms were sterilized. It was the beginning of the unbroken—so far—Goldilocks phase of our planet, at least 3.8 billion years long and counting. How it has kept that way is a mystery at the heart of this book. But first we need to find out how a planet can be gently warmed.

Steps towards greenhouse ideas

The idea of what it takes to keep a planet warm most clearly derives from the work of Joseph Fourier (1768–1830). It was lucky that he and his work survived, as his life was caught up first in the French Revolution and then in the rise and fall of Napoleon Bonaparte. A talented mathematician from his youth, he was unsure whether to devote his life to teaching or to religion. History then stepped in, and he was swept up by politics, and joined the local Revolutionary Committee in Auxerre. It was a move provoked by the hope of supporting this 'greatest and most beautiful cause'. Such idealism, however, was soon confronted with reality in the shape of the Terror that Robespierre had unleashed. Fourier saw this, too late, and wanted to back out—but couldn't. Nevertheless, he tried publicly to defend colleagues of one of many factions in this most faction-ridden of times. It was a dangerous move. Eventually arrested, Fourier perhaps escaped execution only because Robespierre himself fell victim to Madame Guillotine first.

His subsequent career was entangled both with Napoleon, and somewhat reluctantly, with public administration—at Napoleon's behest. Nevertheless, he managed to pursue his interests in the physics of heat, and puzzled at length on how heat was distributed around the Earth. It came, he considered, from a mixture of the Sun's heat, the Earth's own internal heat, and the heat from the stars. The latter, he argued, were very distant but very numerous, and so he estimated the temperature of outer space to be something like the temperatures at the poles—cold enough 'to freeze mercury'. We now know that the stars are very widely scattered in space. Thus, their combined rays can scarcely heat a distant object more than the three degrees above the absolute zero of the cosmic microwave background that permeates the Universe, that almost-cooled echo of the Big Bang that was to be discovered two centuries after Fourier's time.

Nevertheless, in hindsight, Fourier's insights and intuitions are astonishing. He understood that the interior of the Earth contained an immense amount of heat, but that it was conducted to the surface very slowly: so slowly, in fact, that its contribution to the temperature that we feel around us is negligible. He knew that the Earth should be losing this 'primitive heat' (*chaleur primitive*) very, very slowly—and would eventually lose it altogether—*unless there existed other sources of the Earth's heat, yet to be discovered.* This is not to say that he foretold the discovery of how energy from radioactive decay keeps the interior of the Earth hot—but he had the intellectual breadth (and humility) to suspect aloud that, somewhere out there, were forces that lay beyond his calculations.

His place in the history of climate science, though, is based mainly on his insights into how a planet's temperature is controlled by its oceans and atmosphere. He recognized that masses of air and water transported heat around the Earth's surface, and so tended to equalize the temperature at the surface, and he saw that this was achieved mainly by convection rather than conduction. He also saw how air and water are transparent to energy from the Sun arriving in the form of sunlight, but block some of the 'invisible radiation' (*chaleur rayonnante obscure*—which we now know to be long wave-length, infra-red radiation) given off by sun-warmed objects.[8]

How important was this effect? He knew that any heating of the atmosphere by absorption of this 'invisible radiation' would be transported away from the ground as the air, thus warmed, would rise to the higher levels in the atmosphere. So one needed to imagine an

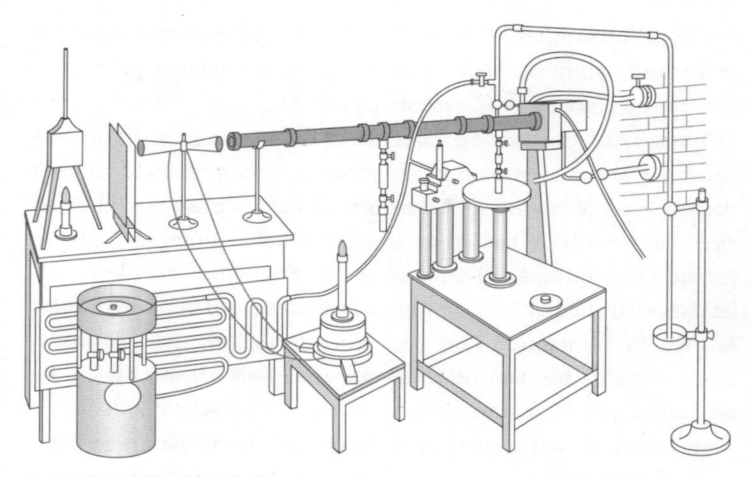

FIG. 4. Tyndall's remarkable device for measuring the heat-trapping properties of gases.

immobile atmosphere—or to simulate one. Here he turned to a device, a solar oven, constructed by that 'celebrated traveller', early alpinist and savant Horace-Bénédict de Saussure. This was a vase, the interior of which had been coated with charcoal to absorb heat, and which had been put inside anything up to three glass boxes, thus trapping layers of air between each of these glass shells. Exposed to the Sun, the air temperature inside the vase could rise to over 100 °C—a temperature that was not affected by that of the air outside. It was not *quite* a greenhouse (and Fourier did not use the term 'greenhouse effect' himself). Nevertheless, Fourier clearly had this concept in mind, and recognized its character and its significance in enabling atmospheres and oceans to regulate temperature, not just on this Earth, but on planets throughout the cosmos.[9]

Fourier did not know what in the air was having this heat-trapping effect, or recognize that this property was determined by a small number of trace gases, which we now call the greenhouse gases. It was John Tyndall (1820–93), a successor and protégé of Faraday at the Royal Institution of Great Britain, who came to recognize the action and significance of the greenhouse gases and to explain what came to be known as the 'greenhouse effect'.

Tyndall had researched the glaciers of the Alps, and in the process became a noted mountain climber, being among the first to reach the summits of both the Matterhorn and the Weisshorn. The glaciers, about which he wrote enthusiastically, prompted him to extend the enquiries of De Saussure and Fourier on how light and heat move through the atmosphere. Tyndall had studied with Robert Bunsen in Germany.[10] This background, and his own curiosity, made of him a brilliant practical experimentalist. The problem he set himself in the late 1850s tested his ingenuity to the full.[11]

He found himself in a virtually unexplored landscape. Today's scientists have to chew through a mountain of published information before getting down to work. In Tyndall's day the literature on how 'radiant heat' and gases interacted could be stated 'in a few words': Macedoine Melloni, with his 'admirable thermo-electric apparatus', could not detect any absorption of radiant heat by atmospheric air, while one Dr Franz of Berlin had detected that atmospheric air absorbed 3.54 per cent of the heat transmitted through his apparatus. Tyndall set out to do better.

What he needed was a tube that he could fill with air, through which to pass the rays of heat. To work out how much of this heat was stopped by the air in the tube, and how much passed through, he needed a measuring device. He used a galvanometer, a silk-wrapped copper wire wound into a coil. Heating it set up an electric current that produced a magnetic field that deflected a needle.

It didn't work. The copper needed to be pure and free of magnetic impurities, so that the needle at rest would not be deflected. This was not the case. The needle took up an initial resting position at a deflection of 30 degrees—useless for Tyndall's purposes. Substituting this coil for another one with fewer impurities brought it down to 3 degrees, but for Tyndall, seeking precise results, even this residual imperfection was a 'source of great annoyance'. He wanted precision and 'almost despaired of obtaining pure copper wire'. He hit upon testing the wire both with and without the green silk coating it came in—and found that copper without the silk was not magnetic! The green dye used for the silk had included traces of iron compounds, the source of the problem. Rewinding the copper wire in pure white silk gave him the precisely poised needle that he needed.

But then he needed somehow to close the tube at both ends—but also to transmit radiant heat through it perfectly. What solid could be as open to the passage of heat, of infra-red radiation, as glass is to light? Rock salt has this property—but to obtain plates of it that were large and transparent enough was terribly difficult. To obtain just a few square inches of absolutely clear rock salt, he had to call on help from a veritable army of the great and the good of learned society.[12] The execution of this single task provides some measure of Tyndall's tenacity—and of his diplomacy.

So now, he had his apparatus, the source of heat rays a container of hot water (he was to try endless configurations too, here, of hot oil, heated copper, molten metal). He started by taking measurements in a vacuum, having pumped the air out of the tube. That gave the baseline. Then, he put in dry air—which gave no detectable result. Then oxygen, and nitrogen, and hydrogen—none of which seemed to shift the needle. They all seemed perfectly transparent to the rays of heat. But, by refining the galvanometer to react more sensitively, he detected some absorption. He carried on working, experimenting for eight to ten hours daily without a break for seven weeks in 1859, resuming in the autumn of 1860, in an 'incessant struggle with experimental difficulties'. It was easy to obtain approximate results—but to obtain precision was *much* harder.

He had succeeded in producing a deflection of about one degree in the needle from dry air ('which had been freed from moisture and carbonic acid'); and the same from oxygen, nitrogen, and hydrogen. But if the oxygen contained a little ozone, its absorption was trebled, the needle swinging to a full three degrees. So, most gases were almost completely transparent to radiant heat—but a few were a little more absorptive than others.

Some gases went far beyond that. Tyndall tried 'olefiant gas' (ethylene) on 21 November 1859. The needle deflected by up to 75 per cent, which equated to absorption of about four-fifths of the total heat. Tyndall was thunderstruck, and his words vividly convey the thrill of discovery: 'Those like myself who have been taught to regard transparent gases as almost perfectly diathermanous (transparent to heat), will probably share the astonishment with which I witnessed the foregoing effects.' To make sure that this astonishing result was not some sort of error, he repeated the experiment with ethylene several hundred times, until he was sure of the result—and even at low pressures this gas continued to absorb most of the heat energy that was passed through it.

He found other gases that absorbed heat, including ether and chloroform. Then he tried 'carbonic acid' (carbon dioxide) and water vapour, and found that these—if not so strikingly absorptive as the hydrocarbon gases—also absorbed a significant part of the 'obscure rays' emanating from warm objects. He realized that these minor, trace components of the atmosphere were its main heat-trapping components, and that varying their proportions 'may have produced all the mutations of climate which the researches of geologists reveal'.

It was a giant step in understanding the controls on planetary climate—a step accomplished by pursuing an idea with great ingenuity, crude materials, and (because of the latter) almost superhuman care and attention to detail. Since then, these 'greenhouse gases' have been augmented by others that have the same property, including those human-produced geological novelties the chlorofluorocarbons (CFCs), while the complex effects of clouds and dust have also been added into the mix of the Earth's heat regulation. Nevertheless, our understanding of climate evolution today is based upon the foundations built by Fourier, Tyndall, and their successors. And yet, as soon as the first real evidence of Earthly climate is preserved, from 3.8 billion years ago, a problem appears.

The mystery of the faint early Sun

The Sun is a star of a common enough sort, ten a penny in the Milky Way, and surely abundant also among the myriad galaxies that lie beyond. The infinitesimal time-slice of its present state can be observed by astronomers. Its past and future, too, can be pieced together by examining similar stars that are within reach of our telescopes and that are at different stages of their life cycle: some just born amid swirling dust-clouds, others in their death throes, and many others, like our Sun, somewhere in the middle of a long life.

In the Hadean, the young Sun was also a faint Sun, emitting about 70 per cent of the light and heat that it does today. About 3.8 billion years ago when the Hadean gave away to the still distant but *slightly* less mysterious world of the early Archaean, the Sun had brightened a little, to some 75 per cent of its present value. And this has become a problem because, in the Archaean, we begin to have direct evidence of the Earth's surface. This surface seems to have been warmer than it should have been, because those earliest rocks preserve no evidence that this was an ice-bound planet. Given, then, that the Sun's rays were much weaker—what wove its thicker blanket?

This question has been debated for the best part of half a century, as the puzzling non-iciness of early Archaean times continued to emerge from the evidence of the rocks. The atmosphere, we know, had little or no oxygen in it, as river-laid strata of that age include grains of minerals such as pyrite and uraninite that, in today's oxygen-rich climate, would quickly oxidize, and rust away.[13] Among the ingredients, many have suggested high levels of carbon dioxide from sources such as volcanic eruptions. However, extremely high levels of this gas would be needed to compensate for the faint early Sun—perhaps at about a hundred

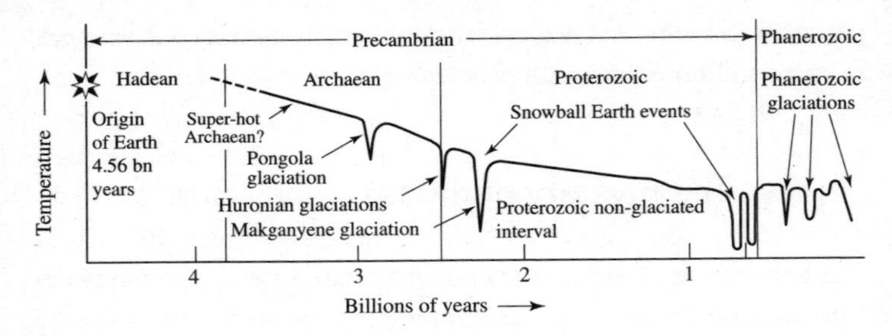

FIG. 5. Earth climate from 4.56 billion years ago to present. The Earth's earliest climate state may, perhaps, have been super-warm. Icehouse climates began during the Archaean. Long-term intervals of greenhouse climate were punctuated by glaciations in the early and late Proterozoic that may have been Snowball Earth events. The glaciations of the Phanerozoic were less extensive, never extending to within 30° of the equator.

times current atmospheric levels. How can we tell if an ancient atmosphere had a lot of carbon dioxide or a little?

One way is to look at strata thought to represent ancient land surfaces—soils. Some minerals, such as siderite (iron carbonate), need high levels of carbon dioxide in order to form. So—if we find siderite in fossil soils[14] can we infer that carbon dioxide levels were high at the time of its formation? Not necessarily, because it depends on what other minerals were there to influence the reactions, and on the micro-environment. Even at a small depth within a soil, the carbon dioxide levels may be very much higher than in the air above, because of the activity of soil bacteria. And even moderately preserved Archaean soils are exceedingly rare. The tentative studies made so far mostly suggest levels twenty or so times higher than today—but not a hundred times higher, and so insufficient to overcome the 'faint young Sun' effect on its own.

So what other additional greenhouse gases might there have been? Methane (CH_4) is a good candidate. This simplest of the hydrocarbon gases is about twenty-five times as potent a greenhouse gas as carbon dioxide. Virtually all methane on Earth today is biologically produced, much of it by the action of microbes called methanogens. It can, however, form without the involvement of life, such as by the reaction of water with certain types of iron- and magnesium-rich igneous rock.[15] In today's atmosphere methane is oxidized to carbon dioxide within a few years. But in a low-oxygen world such as in the Archaean it could have survived for many years in the atmosphere. So that is one possibility. There are other candidates too[16]—ammonia, for instance. This is also a greenhouse gas, but it is easily broken down to nitrogen when exposed to ultraviolet light. Now the best way to keep out ammonia-destroying ultraviolet light is to have a global sunscreen made of ozone: to make ozone a planet needs to have significant amounts of oxygen in its atmosphere, and *that* state was yet a billion years in the future.

Was the Archaean world, perhaps, simply hazy?[17] A haze of organic-rich particles might take on the function of ozone and block out significant amounts of ultraviolet light. The trouble is, it can also block out the warmth-giving sunlight, as the early studies on this possibility indicated. Those early studies, though, simply treated the haze particles as simple spheres. On that mightiest of Saturn's moons, Titan, there is an organic haze—but the particles seem on closer examination to be anything but spheres, rather forming highly complex, fluffy aggregates. Shaped in this way, they are not much of a barrier to incoming light—but they can still absorb ultraviolet radiation. So, if the early Archaean Earth was in that respect, at least, Titan-like, it might have harboured significant amounts of ammonia.

Or the atmosphere of the early Earth might just have been dense. One recent study has calculated that the Earth's crust and mantle contain a couple of atmosphere's-worth of nitrogen. This nitrogen may have once been in the atmosphere[18] before being absorbed by the solid Earth via the mill of plate tectonics. Nitrogen by itself is not a greenhouse gas, but large amounts of it help carbon dioxide and methane absorb more of the outgoing infra-red energy.

Perhaps one should not look just among combinations of greenhouse gases, but consider also the surface of the early Earth. In those days, continents were probably smaller and fewer, because the amount of continental crust seems to have grown through time. Land—particularly barren, unvegetated land—reflects more of the Sun's rays than does seawater; that is, it has a higher *albedo*. If the Archaean Earth was more of a waterworld than it is now, its albedo would also have been lower than it is today, helping it stay warm.[19] Clouds, too, might have been darker than they are today. This is because, with less life than now, there would have been fewer biologically produced particles in the air to act as nuclei to help water droplets condense. If so, clouds then were made of fewer, larger water droplets, and would have been darker, so reflecting less of the incoming radiation.

Peering through the rubble of three billion years to try to find indications of Archaean climate is a little like the challenge scientists faced in trying to understand Mars, before our spacecraft reached that planet. As little as half a century back, the dark patches on this planet could be seriously interpreted as vegetation—vast spreads of lichen, perhaps—from the pattern of their spectra as analysed by the best Earth-based telescopes. And only a hundred years ago, there was not only serious scientific discussion of the possibility of canals on that planet, but speculation that teams of windmills could pump water into their upper reaches, to make and store power for an ancient Martian civilization.

We are similarly groping with patchy, inadequate data to try to make sense of the climate of the early Earth. We know, at least, that the climate could not have been ice-bound. But, even if we also know that the oceans did not boil away—just how hot could it have been, three billion years ago?

A super-hot Archaean?

There have been intriguing hints that the Archaean oceans might have been not just unfrozen, but at times as hot as a freshly poured cup of tea. Part of the evidence comes from those representatives of the modern bacteria that show some of the most 'ancient' characters. A number of these bacteria are 'thermophilic', that is, adapted to high temperatures (up to 80 °C), suggesting that many Archaean bacteria were adapted to these kinds of temperatures.

There is evidence of high temperatures in the rock strata, too, those that are made of silica simply precipitated on to the sea floor to make up layers of the rock known as chert. At formation, the isotopes of both oxygen and silicon, light and heavy, are segregated in different patterns depending on the temperature.The analysis of a number of Archaean cherts has yielded patterns that suggested that the ocean temperatures at times reached as high as 85 °C.[20]

This is a striking result. If true, then it suggests additional means of warming. At those kinds of temperatures, enhanced evaporation of water from the oceans would put a lot of water vapour into the atmosphere, ratcheting up the greenhouse warming. It would also be a barrier to oxygenation of the oceans—both to the introduction of oxygen (no modern photosynthetic organism can survive at temperatures much above 70 °C) and to its retention in the oceans, as oxygen is less soluble in hot than in cold water.

This concept of a super-hot Archaean, however, may be yet another scientific mirage.[21] Such a mirage may be born of isotope values which have been altered to quite misleading values in these old, tattered remnants of the early Precambrian world—or it might stem from a misinterpretation of how isotope systems worked in those ancient and alien oceans. The latest studies (as we write) have examined the pattern of oxygen isotopes in phosphate deposits from southern Africa that are more than three billion years old.[22]

These, by contrast, suggest more modest temperatures, of between 25 and 35 °C.

There is a bewildering tangle of conflicting data here. But it is, we must recall, early days in our exploration of the Archaean climate. This was an eon that was played out over some 1,300 million years. It lasted more than twice as long as the time that separates us from the dawn of the Cambrian Period, when complex multicellular animals appeared in force upon the world. And each sample—so painstakingly collected and analysed for their isotope pattern by scientists—will represent just a snapshot of that enormous time span. Since the Cambrian, the Earth has gone through many climate states, as we shall see in the following pages. The climate of the Archaean will probably have been, in its way, just as much of a rollercoaster as that of the last half-billion years—although perhaps it was a rollercoaster where the highs and the lows were overall from a higher baseline.

The Archaean ended two and a half billion years ago. Towards its end the geology tells us, unequivocally, of a change in the Earth's climate regime. Ice, then, began to get a grip on the world. It came to be a grip so tight that people wondered how our planet could have ever thawed from that state, or how life could have survived. Let us descend towards the chilly depths of Snowball Earth.

Earth as a Snowball

Our attempts to reconstruct the climate of the distant Archaean in Chapter 1 might seem a little like reading a volume of Tolstoy's *War and Peace* recovered from a burnt-out house. Most of the pages have turned to ash, and only some scattered sentences remain on a few charred pages. The Proterozoic Eon that followed began 2.5 billion years ago, thus is not quite so distant from us in time. We know it a little better than the Archaean—at least a handful of pages from its own book have survived. And this book is *long*—the Proterozoic lasted nearly two billion years. This is as long as the Hadean and Archaean together, and not far short of half of Earth's history. Like many a soldier's account of war, it combined long periods of boredom and brief intervals of terror—or their climatic equivalents, at least. The latter included the most intense glaciations that ever spread across the Earth. Some of these may have converted the planet into one giant snowball.

The earliest glaciations

The earliest traces of glaciation on Earth are seen even before the Proterozoic, in rock strata of Archaean age, 2.9 billion years old, near the small South African town of Pongola.[1] These rocks include sedimentary deposits called tillites, which are essentially a jumble of rock fragments embedded in finer sediment. The vivid, old-fashioned term for such deposits is 'boulder clays', while the newer and more formal name is 'till' for a recent deposit and 'tillite' for the hardened, ancient version. Many of the ancient blocks and boulders in the tillites of Pongola are grooved and scratched—a tell-tale sign that they have been dragged along the ground by debris-rich ice. This kind of evidence is among the first ever employed by scientists of the mid-nineteenth century, such as Louis Agassiz and William Buckland, to tell apart ice-transported sediments from superficially similar ones that had formed as boulder-rich slurries when rivers flooded or volcanoes erupted.

Ice, then, appeared on Earth in Archaean times. Was it a thoroughgoing glaciation, though, that spread far across the Earth? For ice can form even in the warmest climates—as today in the snows of Kilimanjaro or the glaciers of the majestic mountains of Irian Jaya in the Far East. In both places the land is tropical, but the mountain peaks rise above 4,000 metres and so are snow-topped and glacier-covered throughout the year (in the case of Kilimanjaro, alas, the ice is rapidly melting). These are simply isolated icecaps on land that is cold because it is high: nevertheless, they can produce those tell-tale boulder-rich till deposits. So, could Pongola have been just the remains of an isolated upland glacier, of little global importance?

Crucially, the Pongola tillites occur together with strata that show signs of having been laid down in a shallow sea. Therefore, the ice must then have more or less reached sea level. So this earliest glaciation was significant—but how extensive was it, and did it extend much beyond the polar regions?

FIG. 6. Ancient glacial till on James Ross Island, Antarctic Peninsula. These sedimentary deposits are formed as glacier ice grinds up and mixes the bedrock it travels across. Tills typically comprise a jumble of rock fragments within a muddy matrix.

Here, magnetic minerals in these ancient sediments provide a clue. These preserve a pattern that records the magnetic field of the Earth at the time the strata were laid down. This 'remnant magnetism' can be used to track the position of ancient continents as they moved across the planet by continental drift. It suggests that South Africa, 2.9 billion years ago, was at a high latitude. So the Pongola glaciation may have been similar to the modest polar glaciations of today, rather than extending towards the equator.

It was in the Proterozoic, though, when ice took a significant hold on the planet. The rock succession of Canada includes tillites from glaciations (the 'Huronian glaciations', they call them) that began some time after 2.45 billion years ago. Their ancient latitude is not well known, but they may have been a prelude to larger events that took place a little later—though 'a little' in terms of the Proterozoic timescale is in this case something of the order of a hundred million years.

These larger events are seen in the rocks of South Africa, to the west of the town of Kimberley, at the edge of the Kalahari Desert. The rocks here are partly covered by the desert sands, but nevertheless show that ice sheets began to move across the region some time after 2.3 billion years ago, before finally retreating some 2.2 billion years ago. These rocks are the Makganyene tillites[2] that lie immediately above a layer of marine strata, suggesting that the ice overrode what had been a sea floor.

The Makganyene rocks are unusual in that, as well as the typical grooved and scratched rock fragments of a tillite, they also contain those microbially formed limestone masses, the stromatolites. These bind together quantities of debris that must have been dropped from melting icebergs that floated above. It is unequivocal evidence that the ice reached the sea. This pattern suggests that the rock debris came in distinct phases, being deposited as the ice advanced, while the stromatolites grew in the intervals when the ice had receded a little. The suggestion of such warmer interludes partly comes from the nature of the stromatolites: these, like all limestones, form more easily in warm than cold water. The mineral that they are made of, calcium carbonate, is unusual in that it dissolves more easily in cold water than in warm water (this is why the great coral reefs occur in warm regions of the world today, and also why they harbour such extravagant, thick-shelled molluscs).

There are volcanic rocks associated with the Makganyene tillites, too, and these preserve remnant magnetism that suggests that this part of the world then lay in the tropics. Thus, if ice reached an equatorial sea in the Proterozoic world, 2.3 billion years ago, then ice, in the Makganyene glaciation, may have advanced across the whole planet, to become the first instance of a Snowball Earth.

At this great age, though, the rock record remains very fragmentary. With further evidence hard to come by, it is difficult to take the Makganyene story further. To demonstrate Snowball Earth conditions properly, one has to show that ice really did cover the Earth, by combining evidence from several former continents. Such evidence can be found much later in the Proterozoic, as we shall see.

There is, though the large question of just *why* the Earth underwent cooling around the end of the Archaean and the beginning of the Proterozoic. Moreover, the Earth was cooling as the Sun was slowly cranking up to produce more heat and light. Something, it seems, was happening to make the Earth leak more heat into outer space. And that 'something', it is suspected, is what ultimately made the Earth just the right planet for *us*. It was the origin of an oxygen-rich atmosphere.[3]

The iron oceans

In the Archaean, the atmosphere contained little, if any, free oxygen. This is clear, as we have noted earlier, from the kind of minerals preserved on the (admittedly few) fossilized land surfaces dating from those times. Nothing rusted. The gases that then formed the atmosphere—an ill-defined mixture of nitrogen, carbon dioxide, methane, ammonia, and such—undoubtedly played a key role in trapping the faint rays of the early Sun.

In the oceans, though, things had started to stir. There, oxygen turns up in a highly distinctive form, in strata in the Archaean, 3.8 billion years ago. The remarkable Banded Iron Formations began to accumulate on the sea floor. These enormous deposits today provide us with almost all the iron that we use. Our cars, aeroplanes, steel girders, cutlery, railway lines—virtually all are made from this

Fig. 7. Banded Iron Formation from the Proterozoic of Zimbabwe. The alternating bands of iron oxides and silica are the result of early microbial activity, as it slowly transformed the chemistry of the Earth's surface.

particular and very ancient source, which exceeds in amount and in purity any other kind of iron ore on Earth.

To the geologist in the field, the Banded Iron Formations look like an enormous, almost endless bar code. The bands consist of thin layers of iron oxide (the bright red mineral hematite) separated by layers of silica-rich sediment known as chert. These distinctive strata were, until recently, widely considered as evidence for early oxygen release by photosynthesis from cyanobacteria ('blue-green algae') as they combined carbon dioxide with water to make more complex organic molecules. The oxygen thus released, it was thought, combined with the giant stores of dissolved iron that had accumulated in the Earth's early oceans, for iron is moderately soluble in oxygen-free water, as anyone who digs a well into a peat bog will discover. Such early 'excretion' of oxygen could have produced insoluble iron oxides that sank to the sea bottom to form bands of iron.[4]

If this is true, the sea must have been highly efficient at absorbing this oxygen—for something like a billion years—before the first undoubted free oxygen began to appear in the atmosphere. The Banded Iron Formations clearly mopped up a lot of oxygen as they slowly formed—perhaps even enough, indeed, to re-fill the Earth's atmosphere to its current oxygen level several times over. While it is true that the oceans are enormous, nevertheless it seems more plausible to have a model for the Banded Iron Formations that did not need such uncannily leak-proof oceans.

In the last few years, an alternative hypothesis has been gathering support. This suggests oxygen-producing cyanobacteria evolved much later than the heyday of the Banded Iron Formations (and much later also than the appearance of the stromatolites). In this scenario the development of Banded Iron Formations in the Archaean and earliest Proterozoic does not depend on oxygen liberated by photosynthesizing cyanobacteria. These iron-rich rocks, rather, could have formed entirely in an anoxic oceanic setting, by the action of microbes which were photosynthetic but which did *not* produce free oxygen—anoxygenic phototrophic bacteria.[5] It's quite a mouthful—yet these bacteria are regarded as the most ancient photosynthetic organisms. They use sunlight, iron, carbon dioxide, and water to make the material of their bodies, and rather than oxygen as a by-product they produce hydrogen and iron oxide. The latter, in this interpretation, made up the brilliant red iron oxide bands in the Banded Iron Formations. In this way, the early oceans can be gradually purged of their iron content without the direct involvement of free oxygen. Whichever mechanism was active, the striking banding of these extraordinary strata may itself be an indication of ancient fluctuations in climate, as the activities of iron-metabolizing bacteria are sensitive to temperature.[6]

The Great Oxygenation Event

Some 2.4 billion years ago, just after the beginning of the Proterozoic Eon, the face of the Earth changed. For over two billion years the lifeless landscapes had been mainly a study in shades of grey and green, wreathed in the mists and organic hazes of those days. Then, the colours changed to shades of red and brown, as the rock surfaces and sediment accumulations rusted. One might wish to have been a painter then. Those rusted sediments can still be seen today, here and there, as layers of 'red-bed' strata.

This was one of the key steps in the mineral evolution of the earth, as literally hundreds of new types of mineral oxide and hydroxide appeared. The 'Great Oxygenation Event',[7] as it is called, seems to mark the origin, within one group of microbes, of the molecular machinery that enabled oxygenic photosynthesis. Seen from the perspective of oxygen-breathing organisms (like us), this was the most fundamental step in the evolution of the biosphere after the origin of life itself. It was a catastrophe, though, for those organisms of the Archaean that had evolved in an oxygen-poor world, and which could not adapt. For these, oxygen was (and remains) a poison, a gas as dangerous and reactive as chlorine is to us. They were gradually forced to the fringes of life in the low-oxygen zones of oceans, or deep into the interior of rocks.

There is further evidence for a low-oxygen atmosphere prior to 2.4 billion years ago, a rather more subtle one found in the sulphur preserved in those ancient rocks. Specifically, it lies in the pattern of isotopes of the sulphur—something that certainly can't be seen with the naked eye, but that requires a good deal of patient sampling and analysis with the atom-counting capabilities of a mass spectrometer to distinguish any pattern. Now, different isotopes of most common elements are generally sifted by a variety of processes. In the case of sulphur, an element with four stable isotopes of different atomic masses (^{36}S, ^{34}S, ^{33}S, and ^{32}S), micro-organisms that use sulphur will tend to select the lighter isotope from whatever sulphur source they are using. That leaves the source richer, therefore, in the heavier isotopes. This process is termed 'mass-dependent fractionation' between the sulphur-source and the extracted sulphur, and the pattern, in terms of the ratios of the light to heavy isotopes, is preserved in the sulphur minerals of ancient rocks.

However, this distinctive pattern is generally not seen in rocks of the Archaean or early Proterozoic. Sulphur preserved in sedimentary rocks older than 2.5 billion years shows, instead, evidence for mass-*independent* fractionation. The ancient sulphur here was released to the atmosphere by volcanic activity as sulphur dioxide. This gas appears to have been subject to a process called photolysis, literally being split by ultraviolet light with wavelengths of less than 200 nanometres. This process is not choosy about which isotopes it selects. Fine sulphur particles, in which the isotopes had not been separated by mass, rained down on the Earth's surface, to become preserved in strata.

The ultraviolet light that is the cause of this kind of fractionation can penetrate the atmosphere only when there is no oxygen. Once oxygen is present, a protective ozone layer forms that acts as a barrier to such radiation. This 'mass-independent fractionation' signature of the Archaean is, therefore, further evidence that the early atmosphere was oxygen-poor. Such photolysis of sulphur dioxide, and the associated pattern of sulphur isotopes, stopped once sufficient oxygen came into the atmosphere.

The history of the oxygenation of the atmosphere is contentious. Some scientific teams have detected large variations in the Archaean and early Proterozoic sulphur record that suggests an atmosphere that may have fluctuated erratically between anoxic and oxygenated states.[8] However, regardless of quite how an oxygen-bearing atmosphere was achieved, it was clearly established by 2.4 billion years ago (while the Banded Iron Formations fizzled out about two billion years ago). Oxygenation had a profound effect on the evolution of life, not least in enabling the formation of an ozone layer that blocks the harmful UV radiation. It is also implicated in changing the Earth's climate, through its effects on other gases in the atmosphere. Its effect upon methane, in particular, may have been key to the refrigeration of the world.

Methane, as John Tyndall's experiments showed (Ch. 1), is a much more powerful greenhouse gas than carbon dioxide—more than twenty-five times as powerful, in fact. It is generally stable in an atmosphere free of oxygen, and can build up to high levels. However, once free oxygen enters the atmosphere, methane rapidly oxidizes to carbon dioxide. Its rapid removal from the atmosphere once oxygen appeared is currently a prime suspect to have tipped the climate into a phase of long-term cooling and initiated the first, severe global glaciation.

The circumstantial evidence for the correlation of the glaciations and of oxygenation is reasonably good. The first of the major glaciations that reached low latitudes, that of the Makganyene diamictites, is approximately synchronous with the main phase of the Great Oxygenation Event. Even the earlier, less extensive ice advances may be consistent with the slender evidence that exists for previous rises in oxygen. That earliest glaciation of Pongola back in Archaean times, for instance, is associated with changes in sulphur isotopes, suggesting that it too might have been caused by an early, short-lived oxygenation of the Earth's atmosphere.

It is early days, still, in the investigation of the ancient cooling phases of the planet. There will be new and ingenious methods developed, no doubt, to interrogate the rather battered and altered rocks that, somewhere within them, contain evidence to support or reject these hypotheses. And, no doubt, there will be surprises too. But even on the evidence we have, the next phase of the Earth's climate is a little counter-intuitive. Surprisingly, after the methane (in quantity) disappeared and ice appeared, the ice did not linger.

The long warmth

After the Makganyene event there is little evidence for a return to global glaciation for nearly 1.5 billion years. This is an enormous span of time—one-third of the age of the Earth—in which global surface temperatures appear to have remained generally stable. It is as though little of significance happened over this time. If that was truly so, we have no idea what mechanism was regulating the climate so effectively, or what prevented further glaciations, especially given the inference that the atmosphere had been largely cleared of methane.[9]

Evidence from rocks, limited though it is, suggests the atmosphere remained oxygenated through this interval, though perhaps oxygen levels were only at about 5 per cent of modern values. The rock succession that follows the Makganyene Formation in South Africa includes deposits with manganese; these too are an indicator of oxygenated oceans—at least in their surface waters (deep waters mostly remained poor in oxygen). Perhaps the Sun's output of heat and light, slowly increasing over this time, was significant in maintaining the long warmth.

Despite the apparent climatic quiescence of the time following the Makganyene glaciation, some important changes were unfolding. For the first two billion years of Earth's biosphere, and indeed right through the Great Oxygenation Event and the Makganyene glaciation, life was limited to simple prokaryotic cells, the ancestors of modern bacteria. But about 1.8 billion years ago this changed with the first fossil evidence of eukaryotic cells—that is, of cells with a nucleus. Shielded from the harmful effects of UV light by ozone, these cells, much more complex than their prokaryotic ancestors, were to open up a new world in ways of feeding and reproducing.

Eukaryotes are thought to have evolved via a process of co-operation, in which a large prokaryote cell probably absorbed a smaller prokaryote. Perhaps these large and small organisms had already been cooperating, making and sharing energy. The smaller absorbed cells were the ancestors of mitochondria (the organelles that produce energy in cells) and chloroplasts (that carry out photosynthesis in plant cells). For most of the Proterozoic, these organisms remained unicellular. Their means of obtaining food became more varied, though, whether through the use of sunlight or not. From one billion years ago, the fossil record shows that eukaryotes were using sex to reproduce, sharing genetic materials, and thus accelerating the process of evolution. These events, although carried out at microscopic level, would ultimately impinge on global climate. As these organisms became larger, more complex and more diverse, they evolved new means of sequestering carbon into their bodies and of cycling that carbon through the biosphere, atmosphere, and hydrosphere. Indeed, the increase in size and complexity seen with the rise of cell cooperation, which ultimately led to multicellular organisms, seem to have coincided with increases in the amount of free oxygen during the Proterozoic. This suggests that there were strong feedbacks between life and its environment.

By about 750 million years ago, the continuing action of photosynthesizing micro-organisms had caused oxygen in the atmosphere to rise to perhaps half modern levels. The ensuing 220 million years of Earth history were to show extraordinary climate change, and an evolutionary jump that would establish most of the animal groups we have today. These are the intervals of time that geologists call the Cryogenian and Ediacaran, the terminal chapters of the Proterozoic. With these, we are entering Earth's maturity. Perhaps appropriately, the late adolescent stage was a stormy one.

Budyko's frozen world

If you have visited the polar regions and looked out, amazed, over the vast ice sheets of Antarctica or Greenland, then you will have a sense of how cold and severe a Snowball Earth, entirely covered with such a frozen carapace, would have been. As far as the eye could see would be blindingly white.

At about the same time that geologists were finding evidence for ancient, very extensive glaciations of the late Proterozoic, the Belarusian climatologist Mikhail Ivanovich Budyko (1920–2001) was at work. Budyko was a survivor—both of the siege of Leningrad and of official Soviet displeasure later in his career. Despite such experiences, his scientific insights were to prove highly influential. He hypothesized that a severe icehouse climate state, in which the whole world was covered in ice, could, theoretically, exist on Earth. But had it done so, he said, then all life on Earth would have been extinguished. As life on this planet evidently exists, he concluded that the Earth has never been fully frozen.

Mikhail Budyko worked at the Leningrad Geophysical Laboratory. From his work in the 1960s and 1970s he is widely regarded as one of the founders of modern climatology, best remembered for his calculations on the amount of the Sun's radiation that is absorbed by the Earth versus the amount reflected back into space—the 'planetary albedo'. Snow and ice reflect light and have a high albedo: the Greenland and Antarctic ice sheets effectively act as giant mirrors, reflecting heat back into space, and this helps to regulate our climate. In contrast, sea water absorbs heat and has a low albedo. Albedo therefore acts as a feedback mechanism, in the case of ice to keep the planet cool, or in the case of open water to warm it. Land types and vegetation cover have intermediate albedo compared to snow or water.

Budyko was a brilliant and original figure[10] who believed that scientists should stick their necks out to defend their ideas. He was, for example, among the first to think about 'global dimming' as a method for preventing global warming. He noticed that major volcanic eruptions were followed by cool weather conditions. The classic example of this is the eruption in 1815 of Mount Tambora, an

Indonesian volcano that threw over a hundred cubic kilometres of ash into the atmosphere, which in the next year had migrated into the stratosphere and caused the 'Year without a summer', crop failures, and widespread hunger in northern Europe. Budyko hypothesized that spreading fine particles into the stratosphere would reflect sunlight back out to space. Indeed, it is still possible to see the effects of those fine particles emitted from the volcanic eruption of Tambora if you look at the paintings of J. M. W. Turner from 1816. The red sky in the background to Pendennis Castle, Cornwall, for instance, is not some whim of Turner's creative genius, but rather shows the effects of the Sun's rays bouncing off the tiny volcanic particles in the upper atmosphere.

One of Budyko's key insights was the importance of feedback mechanisms and their role in maintaining the delicate balance of Earth's climate. Climatologists recognize two different types of feedback mechanism. Positive mechanisms are self-reinforcing: thus, for example, if the climate cools and polar snow and ice cover expands, this will result in more sunlight being reflected back out to space, which in turn will encourage the planet to become colder still. Conversely, if the climate warms and polar ice and snow cover shrinks, the water and land that replace the snow cover would have lower albedo, absorb more heat, and encourage the planet to warm even more. Hence the worry from modern climatology about the recent loss of sea ice in the Arctic: this will act as a positive feedback mechanism in a warming world.

In Earth's climate, positive feedback mechanisms are offset by their negative counterparts, which are self-limiting. For example, if global temperature were to rise rapidly the atmosphere would warm and hold more water vapour. In itself this is a positive feedback (water vapour is a greenhouse gas). But, as that water vapour condensed to form clouds, cloud cover would increase and would block some of the sunlight entering the atmosphere and reflect it back out to space. In this sense the clouds may act as a negative feedback mechanism that tends to limit any rise in global temperature. The physics of clouds is ferociously complicated, and so the relative importance of the negative and positive feedbacks here are still uncertain. Feedback mechanisms are critical to understanding the Snowball Earth hypothesis.

The positive feedback mechanism of snow and ice cover is known as the ice-albedo effect. Budyko calculated that if climate cooled and ice reached to within 30° of the equator, the ice-albedo positive feedback mechanism would become a runaway effect and the Earth would completely freeze over. What Budyko had discovered was a mechanism for producing a Snowball Earth—a world completely covered by ice, both on land and over the oceans.

The evidence of the rocks

What geological evidence is there that Budyko's concept of a wholly frozen Snowball Earth might actually have existed? The 2.4-billion-year-old low-latitude glaciation preserved in the Makganyene rocks of South Africa might be a candidate—but there is just too little evidence from this very distant time to say for sure. More evidence has been preserved from the later part of the Proterozoic, though, in rock strata that are somewhere between 740 and 580 million years old. And here clearer indications of a deep-frozen Earth have emerged.

Deposits of these late Proterozoic glaciations were first recorded in the late nineteenth century. Suggestions that they may represent a global Proterozoic glaciation were first made in the 1930s and 1940s by Douglas Mawson (1882–1958), the Australian (though Yorkshire-born) geologist. Mawson is more generally known as one of the great figures of the 'heroic age' of Antarctic exploration. His expedition to the then almost completely unexplored King George V Land was a true epic, and a tragic one. Of the three people on the expedition,

Mawson was the sole survivor: one colleague plunged into a crevasse—together with most of the food supplies and the best dogs—while the other colleague died later, probably of an overdose of vitamin A (they were forced to eat the sled dogs, the livers of which contain very high levels of this vitamin). Mawson, now alone, survived another crevasse-fall and made it back to base camp—but only a matter of hours after the ship, with most of the men, had sailed away. He had to overwinter there, in conditions of almost permanent blizzard, with the six men who had remained in hope of his return.

Later, working in a less hostile landscape, Mawson recognized rocks of glacial origin in Australia that lay beneath—and were hence older than—strata that contained some of the world's oldest fossils of multicellular organisms. Mawson believed that the record of glacial deposits in Australia, and of similar deposits in Africa, indicated an ancient global glaciation that reached the tropics. This was prescient—but, at the time, not wholly convincing. Mawson did not believe in plate tectonic theory and hence thought that Africa and Australia had always occupied their present positions on the Earth. As a result, his idea did not win wide approval. Other geologists explained the glacial deposits as having formed as these continents drifted through the polar regions at different times.

It was another geologist, Brian Harland (1917–2003), who really set the snowball rolling. There were similarities between the two men: Harland was also originally from Yorkshire, from the seaside town of Scarborough, and was, in his youth, inspired by its famous geological coast. By the age of 13 he had already uncovered the bones of an ancient Jurassic crocodile. He too, like Mawson, became a renowned Arctic explorer, leading twenty-nine expeditions in the days when this kind of exploration was still essentially as it was in Mawson's time—heroic, romantic, and dirty. Apart from a brief time teaching in Chengdu, China during the 1940s, Harland spent his academic career at Cambridge, though he was always a prodigious visitor of distant lands.

His journey into the long-vanished ice of the late Proterozoic began with evidence published in 1959, describing preserved signals of the Earth's magnetic field during the late Proterozoic, from ancient glacial deposits in central Norway. Now at high latitudes, these rocks were deposited from ice at an ancient 'palaeolatitude' of 11°.

Glacial deposits of late Proterozoic age are now recognized worldwide, from Arctic Canada to Australia.[11] Three separate glaciations have been recognized: the Sturtian, beginning about 740 million years ago; the Marinoan from 660 to 635 million years ago; and a less severe event, the Gaskiers glaciation, about 580 million years ago. In any one region there may be one, two, or sometimes all three of these glacial episodes represented. These events seem to have been synchronous around the Earth—although as not all the glacial deposits are easy to date radiometrically, this in itself has been a topic of intense (and at times furious) debate. The time encompassing the Sturtian and Marinoan glaciations is now talked about as representing a distinct period of geological time, evocatively termed the Cryogenian; it is still informal, but looks set to be formalized as a geological period once this particular episode of Earth history has been dissected, charted, and analysed in more detail.

The detail is rapidly piling up: in the last couple of decades there has been a resurgence of interest in the Snowball Earth hypothesis driven by scientists such as Joe Kirschvink from Caltech and Paul Hoffman from Harvard. We know, for example, that the Marinoan glaciation must have been long-lived. The alignment of magnetic minerals in the late Proterozoic glacial deposits of Western Australia records evidence for a number of reversals of the Earth's magnetic field, indicating that the ice endured for millions of years. Some of the glacial deposits evidently formed at high latitudes, though at many sites in Africa, North America, and Australia their preserved magnetism shows that they were laid down in the tropics—some of these, at least, in a marine setting. This was a world of ice, for sure. How, then, did the Earth freeze over?

Planet of ice

Mikhail Budyko's model assumes that the positive feedback mechanism of the ice-albedo effect would, once ice grew to within 30° of the equator, be unstoppable, and the whole planet would freeze over. But for the ice-albedo effect to operate, ice must first grow: there is a chicken before the egg.

We have already seen that the liberation of free oxygen to the atmosphere during the early Proterozoic would have oxidized methane to carbon dioxide and hence reduced greenhouse warming by the atmosphere. Coupled with solar luminosity at perhaps 85 per cent of modern levels, this may have driven the Earth into the severe glacial state of the Makganyene world. But by 750 million years ago, free oxygen had been accumulating in the atmosphere for at least 1.4 billion years (and perhaps longer) and may have approached levels of about 50 per cent of those of today. In tandem, any methane released to the atmosphere was rapidly oxidized to carbon dioxide: by the late Proterozoic, carbon dioxide was the more significant atmospheric greenhouse gas.

What then, triggered the intense glaciation of the late Proterozoic? Was it plate tectonics? Evidence from remnant magnetism preserved in Proterozoic rocks indicates that at first, the small, early continents wandered across the planet rapidly, 'floating' and being driven by convection in a mantle that was hotter than today by up to a couple of hundred degrees Celsius, and therefore circulated more quickly. As time went by this process slowed, and more and more 'light' granitic continental crust accumulated and accreted. By the late Proterozoic there were substantial areas of continental crust, built largely of granitic materials and the sedimentary rocks derived from them. The movement of continents by plate tectonic drift was probably a major influence on long-term climate change. An expression of this influence may have been the break-up of a supercontinent known as Rodinia, about 800 million years ago.[12]

The break-up of Rodinia—the 'homeland'—left many smaller continents in a palaeo-equatorial setting. In such settings, rocks are more easily chemically weathered by tropical storms and rainfall than in the temperate or polar regions of the Earth. When rain falls through the atmosphere it dissolves carbon dioxide to form carbonic acid [H_2CO_3]. The carbonic acid can then react with silicate minerals such as feldspar—one of the kinds of mineral that form granitic continental crust. The weathering reactions produce calcium and bicarbonate ions, which are carried to the oceans by rivers. Once in the sea, the calcium and bicarbonate ions are fixed together by marine organisms to form limestone. In effect, carbon is scrubbed from the atmosphere and buried in the sea.

The process of reaction of carbonic acid with silicate minerals is called 'silicate weathering' and it operates as a powerful negative feedback mechanism upon climate (see also Chs. 3 and 5). With more carbon dioxide in the atmosphere, global temperatures rise higher—and as temperatures rise, more water vapour is evaporated from the oceans into the atmosphere. This evaporated water eventually falls as rain, mixing with the carbon dioxide to form carbonic acid and so enhancing the silicate weathering. Therefore, as global temperature rises, silicate weathering works to counter the effect by removing carbon dioxide. (This process works slowly, over timescales of hundreds of thousands of years, so the counter-effect is too slow to prevent the global warming predicted for our century.) When global temperature is too low, the converse effect works, as there is less rainfall and thus slower silicate weathering. This appears to be a key process that, overall, has protected and maintained our Goldilocks planet. Perhaps it is *the* key process (though one should not forget the importance of the Earth's living organisms, too, in the regulation of atmosphere and climate).[13]

Rodinia, as it broke up in its tropical setting, may therefore have acted as a kind of a giant silicate-weathering factory. As carbon dioxide was removed from the atmosphere, glaciation ensued. Once ice grew to within 30° of the equator, the ice-albedo feedback kicked in, and the ice spread to cover the Earth. In the 'purist' view of this process, espoused by Paul Hoffman, Joe Kirschvink, and their colleagues, the Snowball was complete, with ice spreading to cover not only land but also the entire sea surface.[14] This is a vision like Antarctica in its winter—but worldwide. It is a frozen world, and a blindingly white one, as the sunlight is reflected from its surface—except for scattered mountain nunataks that would have poked up out of the ice, here and there. It would also have been a hyper-arid world. The hydrological cycle would have stopped, as water could no longer be liberated from the frozen ocean surface, and so there would have been no clouds, no rain—and no snow either. Erosion and silicate weathering, therefore, would also have stopped.

As the ice thickened, up to a kilometre in places, global mean temperatures would have plummeted to perhaps −50 °C (it is +15 °C today). Light penetration, already weakened at the surface of the oceans, would be very low. As photosynthesis slowed, and then largely ground to a halt, the oceans would have become mostly anoxic as the oxygen was used up. The long-forgotten Banded Iron Formations briefly reappeared, for the first time in a billion years, perhaps because oxygen starvation in an ice-covered ocean once more allowed iron levels to build up in deep ocean waters. This is a planet that to us, today, would be quite unrecognizable. But was it real? Did such a scenario occur?

Other scientists have taken the same evidence and interpreted it differently, to conjure up different types of Snowball Earth. The contemplation of alternative worlds is one of the privileges of a geologist, after all.

Alternative worlds

The Snowball Earth model represents one vision of the Sturtian and the Marinoan glaciations. As we write, this is highly controversial territory. These Proterozoic glaciations were certainly extraordinary, but by no means all scientists agree that the whole of the Earth's surface, land and sea, was then ice-bound. There are alternatives to the Snowball (even when it has as passionate and eloquent an advocate as Paul Hoffmann). These alternatives have wonderful names, such as 'slushball', 'zipper rift', and 'high tilt' Earth.[15] None of these theories doubt the fact of extremely cold conditions on Earth in the late Proterozoic, but they differ in scale.

Chief amongst the protagonists is Slushball Earth, an idea closer to what Brian Harland was originally thinking of when he speculated on a global glaciation in the late Proterozoic. In the Slushball hypothesis, glaciation finds its way into the tropics, but areas of sea remain permanently open—so some form of hydrological cycle can persist, and carbon dioxide can also exchange between atmosphere and sea water. The arguments for a slushball rely heavily on liquid water, inferred from glacial deposits that grade laterally into water-lain deposits, which are then regarded as indicating areas of open sea.

A more radical interpretation is the zipper-rift hypothesis, championed by Nick Eyles of the University of Toronto. In this hypothesis the break-up of the Rodinia supercontinent produced mountain ranges that, on being eroded, poured sediment into adjacent areas of lower ground. In zipper-rift, glaciation takes place on those scattered patches of mountainous terrain. Zipper-rift does not require a global glaciation.

Both slushball and zipper-rift are testable hypotheses as more data accumulate from the geological record of the Proterozoic. Firm evidence for the movement of materials by near-surface liquid water would support a slushball hypothesis, while a clear demonstration that many of the glaciations took place in fault-bounded rift basins, together with a precise chronology to show that the glacial deposits were *not* of the same age everywhere, could be used to support the zipper-rift model.

Perhaps the most radical interpretation of late Proterozic glaciation is the concept of a high-tilt Earth. As the Earth orbits the Sun once a year it is continually spinning on its axis, once every day. Earth's axis of rotation is inclined at between 22.1 and 24.5 degrees from the vertical, probably as a result of the collision with the short-lived planet Theia early in its history (see Ch. 1). George Williams of the University of Adelaide came up with the radical idea that, during the late Proterozoic, the Earth's axis of rotation may have been tilted at a much greater angle, perhaps by as much as 54 degrees. Then, the polar regions would receive more heat than the tropics, and this would explain why so many late Proterozoic glacial deposits are preserved at equatorial latitudes. There is a problem with this model too, not least that the Earth's movement through the heavens is coupled to the Moon, whose gravity effectively constrains its range of manoeuvre. Proponents of the model have got around this by suggesting that the Earth's crust and mantle may have shifted en masse relative to the core. Intellectually, this is a time of high adventure in the late Proterozoic. As yet, there is no clear winner among the competing ideas (though proponents of the Slushball currently seem most numerous). The discussion seems set to run for some time yet.

A model snowball

Given the fierce debates about what a Snowball Earth was really like, some scientists have attempted to resolve the issues by building complex computer models to try to simulate such a world.

These models have been evolving since the 1960s, and their complexity has been rising as computers have grown more powerful. Some of the most advanced computer models can simulate the whole Earth, like those of the Hadley Centre computer of the UK Meteorological Office. These are the kinds of supercomputers that predict our weather, and they can carry out fifteen billion calculations a second: this kind of power is needed to model something as complex and dynamic as the Earth's atmosphere.

The models are usually referred to as General Circulation Models, or GCMs for short. GCMs divide the Earth into a series of grid boxes. Within each of these boxes, variables important for the prediction of climate are calculated, based upon the laws of thermodynamics and Newton's laws of motion. The reactions between the individual grid boxes are calculated in progressive time steps. All such climate models of ancient Earth have one thing in common: they must be grounded in real geological evidence. It is this evidence that provides the *boundary conditions* for the model. The boundary conditions might include atmospheric greenhouse gases, sea surface temperatures, the altitude of mountain belts, and the distribution of ancient ice. All these boundary conditions can be calculated, or estimated, from geological evidence, though the further back in time we go the more difficult this is to do.

Climate modelling is a little like a complex, multidimensional chess game. Each of the grid boxes can be viewed as a single chessboard, and the variables within it as chess pieces. The chess pieces are free to move only in certain ways and in certain directions. Given so many possible outcomes, rules are needed. These rules are set by the boundary conditions, which define the nature of the world on which the game is set. When the game begins, what takes place on one chessboard can affect what happens on another. The modellers, those grandmasters of this version of climate chess, look on to see what the outcome will be—and ponder on just how closely the game matches reality.

Among the grandmasters of this particular game are Mark Chandler and his colleague Linda Sohl of the NASA Goddard Institute for Space Science at Colombia University, New York. Mark is a sedimentologist and computer whiz, while Linda is a geologist who specializes in the Proterozoic. Between them they have the kinds of skills needed to model the Snowball Earth: a solid grounding in the geology of this ancient time, and knowledge of supercomputers and climate models.

To build a model of the late Proterozoic Earth,[16] Mark and Linda had first to identify the boundary conditions of such a world: to specify such features as the surface albedo, the atmospheric composition, the position of the continents, the way the oceans transfer heat from the tropics to the poles, the amount of sunlight, and its distribution. All these boundary conditions have a major influence on how Earth's climate functions—both in reality and in the model. The hope is that the models, as they are developed and made more sophisticated, will behave more and more like the real Earth.

This far back, 700 million years ago, most of the boundary conditions are not easy to set. Only the amount of sunlight is easy to fix, as this follows a path that can be reconstructed based on the conversion of hydrogen to helium in the Sun's interior. It follows that the Sun's energy output was 94 per cent of present values then.

Other boundary conditions, such as the positions of the continents, were more difficult. Palaeomagnetic data gives evidence only of latitude, not longitude—and even that evidence is patchy. Similarly, the levels of greenhouse gases are poorly known.

Having fixed their parameters, Linda and Mark ran the model with modern boundary conditions to check that it would simulate the modern climate. All climatologists do this to test how well their models work. They then ran the Proterozoic model for a simulated 100 years, so that the model could reach an equilibrium state according to the given parameters.

What happened? Did snow extend to the tropics and cover the whole Earth in ice? Well... actually, no. What Mark and Linda found was that the model would produce an ice age rather like our own, with snow and ice at high latitudes, but with warmer conditions persisting closer to the equator. Even at the lowest atmospheric carbon dioxide level of 40 ppmv (lower than we think it has ever been, in fact) and the ocean heat transfer reduced to just 10 per cent, they still could not force a Snowball Earth. Indeed, reducing the ocean heat transfer to just 10 per cent effectively keeps all the heat in the tropics, preventing rain falling as snow there to form ice sheets. Even in this worst climate scenario, mean global temperature was only lowered to 3.36 °C—compared to about 15 °C at present, and was certainly much warmer than the −50 °C envisaged for a severe Snowball. Furthermore, ice grows only to within 40° latitude of the equator in the model, so there seemed to be no way of forming Budyko's unstoppable ice-albedo effect.

Models are just models, and the parameters for Linda and Mark's models represent only certain scenarios. The models do show that the very ancient world—and *especially* some of its more dramatic faces—still remains deeply mysterious. While some of the features of the late Proterozoic world are lost to us forever, we can still search the geological record for better clues to how much carbon dioxide the atmosphere might have contained, or where continents were positioned. As we move towards the present, the boundary conditions of climate models become ever more firmly founded. Scientists, though, will keep searching for answers to the riddles of the distant past. They represent, after all, one of the last great adventures.

Underneath the ice

Mikhail Budyko speculated that Earth could never have actually frozen over, for otherwise all life on planet Earth would have perished. Brian Harland had shown that a Snowball Earth seemed possible, based on the geological evidence, and Paul Hoffman and colleagues have spent careers collecting the data to substantiate it. If a Snowball Earth did happen, then why was life not extinguished? In fact, the evidence from the time just after the Snowball glaciations, from the late Proterozoic and earliest Cambrian, shows life evolving at a pace never seen before or since. Where then, did life cling on?

There are places in our Solar System where icy worlds, like the possible Snowball of ancient Earth, persist today. Titan, the largest moon of Saturn for example, possesses an icy outer shell that hides a possible ocean of liquid water below. Scientists have speculated that life may exist there. Europa, a moon of Jupiter, also seems to hide an ocean of salty water below a covering of fractured ice. But there are environments much closer to home where we might get a glimpse of life persisting in the most hostile, refrigerated state of this Earth.

Where on our planet are there environments that sustain life, but that might mimic the conditions of a Snowball Earth? If temperatures were really –50 °C on Snowball Earth then there is really only one place to compare it with: the Antarctic. On that continent, among the most severe environments with life are the lakes of the McMurdo Dry Valleys in the East Antarctic. The Dry Valleys are among the most inhospitable places on the surface of Earth. During the summer months temperatures here average a consistently freezing –12 °C, while average winter temperatures plummet to –25 °C. Worse still, this is also an extremely arid place—one of the most extreme deserts on Earth, in that aspect also mimicking a Snowball Earth. Yet even here, life hangs on.

In the Dry Valleys there are lakes covered by ice up to 8 metres thick, but below which there is liquid water. The water is rich in mineral salts, derived from the weathering of the Antarctic landscape by glaciers; sometimes it may be more than seven times saltier than sea water. In these lakes, the dense, saline deep waters do not mix with the surface waters and therefore receive almost no oxygen.

But even in the cold, dense, bottom waters with little oxygen, there is life.

The organisms of these lakes form simple ecosystems. Not only are the physical conditions harsh, but the ecosystem must also cope with little or no light during the austral winter when the Sun dips below the horizon. There are no fish. Such an environment sustains but a few, miniscule rotifers and the occasional tiny crustacean such as a copepod, living off the proceeds of a food web that also comprises single-celled motile organisms, photosynthesizing phytoplankton, and bacteria. What is so remarkable, though, is that even in the dimly lit, ice-covered waters, phytoplankton flourish and make their food by using photosynthetic systems specifically adapted for shade.

If life can continue in such an Antarctic lake, then Mikhail Budyko's contention that a Snowball Earth (let alone a Slushball) would destroy all life on Earth may not be correct. Yet somewhere, somehow, life clung on, perhaps in the cracks between the ice, or in warm, nutrient-rich waters deep under the sea at the mid-ocean ridges. There then followed an explosion of new life. That, though, was after the thaw.

The limestone connection

There is a puzzling, and seemingly quite incongruous, aspect of the Snowball Earth glaciations. World-encompassing glacial severity is not everything—the Earth's chemistry was undergoing remarkable changes then, too. Even in the earliest days of study, it was seen that the glacial deposits lay cheek by jowl with limestone strata—deposits that are most typical of *warm* temperatures. This seemed strange. Was there some odd combination of extreme cold and balmy warmth—or were these some sort of cold-water variety of limestones? The pattern fell into place a little when the sequence of events was established: the limestones always lay above—and therefore immediately post-dated—the glacial deposits. They were therefore christened 'cap carbonates' and recognized as representing in some way the recovery—the warmth, that is—that followed upon the retreat of ice. And, given that these are carbonates, it was likely that the greenhouse gas carbon dioxide was somehow implicated in their formation. They are part of a planetary escape mechanism from the all-embracing cold, in some way. Such a Houdini-like trick has been envisioned, with the help of Earthly fire.

Escape from the planet of ice

The world as total whiteness, the oceans frozen, and the land covered in ice: Budyko's thermodynamic model suggests that such a Snowball should last forever. As the sunlight is reflected back out to space there is nothing to heat the surface. But if there was such a 'hard Snowball' Earth, we know that this cannot be the case. We are here today, after all—evidence that a planet can return to comfortable and life-enhancing warmth.

There is a way back from this frozen world, an escape hatch opened via the action of plate tectonics. For, on Earth, volcanoes will continue to erupt, even beneath a shell of ice—and even a modest eruption may melt through an ice layer like a blowtorch, for ash and gases to burst out into the atmosphere. On a 'hard Snowball' Earth, though, the hydrological cycle has stopped as the water at the surface is locked in ice, and there is no water in the atmosphere to scrub out carbon dioxide. No carbonic acid will form for silicate weathering. In any case, the land is covered with ice so there are scarcely any silicate minerals at the surface to be weathered. Glacial conditions may persist thus for millions of years. But, during this time volcanoes will continue to pour out carbon dioxide—and also ash, to drape over and darken the icy surface. Gradually, over millions of years, the carbon dioxide in that arid atmosphere can build up to levels perhaps 100 times or more above our present levels. At some critical threshold the atmosphere will become so warm through the action of this greenhouse gas (and perhaps, arguably, of methane too) that it overcomes the ice-albedo effect. Then, there ensues a catastrophic event to rank with any in Earth's history: a sudden, catastrophic melting of a global ice cover.

Imagine: an icesheet as thick as the Antarctic, covering the whole Earth and melting in, perhaps, some hundreds of years—a million Niagara Falls all flowing at once. This extraordinarily rapid melting of ice, the newly kick-started hydrosphere, including the rebirth of rain, and the highly acid nature of that rain—as it began to wash the carbon dioxide out of the air—flushed an enormous quantity of dissolved chemicals into the oceans. This seeded the precipitation of large quantities of carbonate sediment on to the sea floor, bringing about the sudden global change from glacial tillites to cap carbonates.[17]

The melting of the ice seems to have generated intense storms. Sedimentary structures that look like teepees stacked one on top of another are found in the cap carbonates and appear to signal intense winds and waves generated by the catastrophic thaw of Snowball Earth. Paul Hoffman—one of the great exponents of the Snowball Earth hypothesis—and his colleague Philip Allen, then at Zurich, have described these 'teepees' as giant wave ripples, moulded on the sea floor during intense storms.[18] The term 'giant', here, means about 40 centimetres high and about 2 metres long, several times as big as the typical wave ripples in shallow water.

These Precambrian ripples suggest consistently high-speed winds, to produce large and slowly oscillating waves. They are envisaged as having formed during the dramatic thaw, when there was a large thermal gradient between those areas where ice persisted and those where the ice had already slipped away. A thaw—particularly one on this scale—does not always bring kinder weather. When the last of the Snowballs had melted, though, the world had changed. New life was soon to come.

Dawning of a new age

All of us carry molecular clocks inside us. These clocks chart the rate at which our genes have evolved through time. They tell us, for example, that the last common ancestor of humans and chimps lived some eight million years ago. Molecular clocks track the possible dates when different animal, plant, and fungal groups arose. But there is a disconcerting convergence here. The molecular clocks say that the origins of many animal groups occurred in the late Proterozoic, probably at around the time of glaciation 650 million years ago. If that is the case, and the timing of the geological and biological events is not yet precisely constrained, then far from being an ecological disaster, the late Proterozoic glaciations may even have been a spur to evolution.

The nature of any link between these glaciations and the flowering of multicellular life remains highly speculative. There are tantalizing scraps of evidence, though. For instance, there is an indication in the strata of a flush of phosphorus, glacially eroded from land into the Snowball Earth oceans.[19] After the thaw, such super-fertilized oceans would have helped algal microplankton thrive. Drawing carbon dioxide out of the air, these could grow, die, and be buried in larger-than-usual amounts. Their buried remains would take the carbon with them, leaving the oxygen to build up in the atmosphere. And, as oxygen levels rose, the stage would be set for the kind of metabolism necessary to support complex life.

It's a plausible scenario of how climate and life may have interacted in those days (though likely to be far from the last word on the matter). By whatever means, though, in the interval following the last and mildest of the Snowball Earth events, the Gaskiers glaciation of 580 million years ago, enigmatic multicellular organisms called the Ediacaran biota first appeared on Earth. Over the next sixty million years, the fossil record shows that nearly all the modern animal groups appeared, including the first vertebrates, in the form of tiny fish preserved in the ancient rocks of southern China. This change in the complexity of Earth's biosphere heralded the Phanerozoic Eon, the past 542 million years of Earth time (that we still live in). It established new and complex interactions between the Earth system and the biosphere that have persisted to the present. The abundant fossilized remains of the new life-forms allow the history of Earth, and of its climate, to be charted in ever increasing detail. The Snowball glaciations never came back. The last half billion years, instead, show a world—our more familiar world—oscillating between greenhouse and icehouse states.

Between Greenhouse and Icehouse

The battle of Carnival and Lent

There is a celebrated Flemish painting by Pieter Bruegel the Elder in the Kunsthistorisches Museum in Vienna. It depicts the age-old battle between Carnival and Lent. Carnival—a time of high spirits, led in this vision by a fat man on a beer-barrel, carousing and brandishing a pig's head on a spit—is opposed by Lent, deflating the happy excitement and bringing in a time of sobriety and abstinence. Bruegel's understanding of these opposed rhythms of rural life in the sixteenth-century Netherlands was acute: he was nicknamed 'Peasant Bruegel' for his habit of dressing like the local people, to mingle unnoticed with the crowds, all the better to observe their lives and activities.

Bruegel's vision of the age-old rhythm of life, in the form of an eternal oscillation between two opposing modes, may be taken to a wider stage. From the late Archaean to the end of the Proterozoic, the Earth has alternated between two climate modes. Long episodes of what may be regarded as rather dull stability, best exemplified by what some scientists refer to as the 'boring billion' of the mid-Proterozoic, are punctuated by the briefer, though more satisfyingly dramatic, glacial events. This alternation of Earth states persisted into the last half-billion years of this planet's history—that is, into the current eon, the Phanerozoic.

If anything, the pattern became more pronounced, as if it had become an integral part of the Earth's slowly moving clockwork. There were three main Phanerozoic glaciations—or more precisely, there were three intervals of time when the world possessed large amounts of ice—though in each of these, the ice waxed and waned in a rather complex fashion, and none came close to a Snowball-like state. Thus, these intervals often now tend to be called 'icehouse states' rather than glaciations per se. Between these, there were rather longer intervals—greenhouse states—in which the world was considerably warmer; though again, this warmth was variable, and at times modest amounts of polar ice could form.

Of the Earth's Phanerozoic icehouse states, two are in the Palaeozoic Era: one, now termed the 'Early Palaeozoic Icehouse' centred on the boundary between the Ordovician and Silurian periods, peaking some 440 million years ago; and a later one centred on the Carboniferous and early Permian periods, 325 to 280 million years ago. The Mesozoic Era was essentially a greenhouse state, being almost

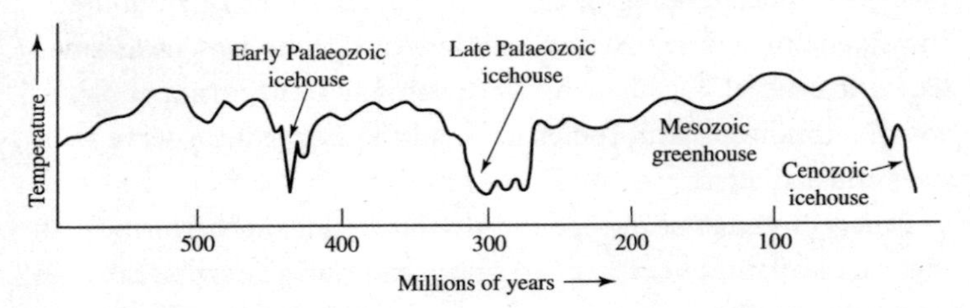

Fig. 8. Phanerozoic time has been characterized by the interplay between greenhouse climate and three great icehouse intervals, during the early Palaeozoic, the late Palaeozoic, and the Cenozoic. We still live in the last of these.

200 million years of virtually unbroken warmth, in which the dinosaurs thrived. We live today in an icehouse world: it began, rather abruptly, mid-way through the current Cenozoic Era, some thirty-three million years ago, and for the last three million years has been intensifying. Its future evolution is, naturally, a topic of great interest to us.

Each of these icehouse worlds of the Phanerozoic has similarities: in each, a large continental ice sheet formed at high southern latitudes, and sometimes one in the north too. Each icehouse shows multiple intervals of intense cold separated by warmer phases. Ice never approached the equator, though: Budyko's vision of a Snowball Earth catastrophe has not (yet) come to pass in this eon. There was another way, too, in which the icehouse worlds of the Phanerozoic and the Proterozoic differed. There was no repeat of the ancient pattern of glacial tillites surmounted by cap carbonates. So, as well as some greater regularity in the Earth's long-term climate behaviour, there is a certain moderation also. Temperatures rose and fell, but not so drastically, and without such great chemical perturbations in the oceans. The Earth, in middle age, seems to have mellowed.

The reasons for this change in pattern are obscure. But it is coincident with—and may be related to—yet another major step in the evolution of the Earth. The Phanerozoic world began with the entrance of a biosphere transformed. For the preceding three billion years the Earth had been, in biological terms, essentially a slime-world of assorted microbes. After this long reign, there was a brief interval of a few tens of million years when the strange Ediacaran organisms spent their enigmatic but probably rather dull lives clinging on to the late Proterozoic microbial mats. Then—revolution!—and a cornucopia of multicellular animals appeared, digging, scratching, searching for food, eating and being eaten. Many of them, too, developed the biochemical trick of building hard skeletons—of calcium carbonate or calcium phosphate. This was largely a means of helping them to predate or to avoid predation in the new Phanerozoic arms race. As a bonus (to us) it rendered their remains fossilizable, and easily visible in the strata.

There are implications for climate in this outburst of complex life: practically, in helping us assemble a climate history; and—rather more subtly—in affecting the course of global climate itself. To start with though, it provided the means to define the world better.

The patterns of complex life

> I cannot doubt that all the Silurian trilobites have descended from some one crustacean, which must have lived long before the Silurian age and which probably differed greatly from any known animal.... Consequently, if my theory be true, it is indisputable that before the lowest Silurian stratum was deposited, long periods elapsed, as long as, or probably far longer than, the whole interval from the Silurian age to the present day.

The singular event in Earth history that Charles Darwin referred to in *The Origin of Species* is the appearance of those fossils of complex animals in rocks some half a billion years old that we now call Cambrian in age (the relationship between the Cambrian and Silurian periods was not, in the 1850s, yet resolved, nor had an intervening Ordovician Period yet been placed between them to separate two academically warring factions). Darwin pondered the sudden, nigh-miraculous, and global appearance of fossils in the rock record at the base of the Cambrian, and he surmised, quite correctly, that the origins of the organisms that left those fossils must lie yet deeper.

There is still considerable mystery surrounding this event; but the complex organisms that emerged clearly showed preferences for different conditions of heat or cold, of humidity or aridity, of abundance or modesty in nutrient levels. Different communities could be separated by barriers—or merge when those barriers broke down. They adapted to their habitats, or migrated to seek out more favourable ones.

Fossils, therefore, provide a new means of tracking the movement of continents, or the different regions of the oceans, or the movement of climate zones, back and forth across the globe. Coral reefs, for instance, and the myriad organisms that shelter in them, are communities of the tropics and subtropics. Towards the poles, these rich assemblages disappear from shallow seas, to be replaced by poorer communities where molluscs and starfish figure largely. On land, diverse tropical forests occur over the equator, while less diverse pine and birch forests extend over more northern climes. These communities, when fossilized, will retain their signature even when the landmass has changed its position utterly. Fossilized tropical forests for instance, can be found today under the ice in Antarctica. Such fossil assemblages are guides to ancient place and ancient climate—guides that did not exist prior to the Phanerozoic.

Further distinctions can be made. Hence, today penguins are found only in the south, along the coasts of southern Africa, Patagonia, and the Antarctic (in one case they have migrated northwards to the Galapagos Islands in the tropics—but this was only with the help of a cold-water, north-flowing Peruvian current). Walruses, though, live in the polar north. The two remain separate, and penguins and walruses only meet in zoos and in children's books. A thermal divide—the tropics—exists between Arctic walruses and Antarctic penguins. So when a palaeontologist looks at the fossils of penguins and walruses 100 million years hence, he, she, or it will discern a remnant of the bipolar glacial world we live in today.

The walruses and penguins are components of northern and southern polar *biotopes*. Biotopes are simply a collection of animals and plants that cohabit and interact, and which define a particular set of environmental parameters. We cannot easily recognize biotopes in the Archaean and Proterozoic fossils, because life was too simple and too remote from us to recognize any sensible analogues. But we can identify biotopes preserved in rocks from the Cambrian onwards. These ancient biotopes moved to and fro over the globe as, say, the ice sheets grew and melted, signalling changes in environment at different latitudes. But fossils also provide another profoundly important tool that enables us to reconstruct the course of ancient climate. They provide us with the ability to establish a precise order of events.

The earliest geologists saw the problem. They were confronted with the enormity of the geological record and could then see no easy means of assembling this gigantic and tangled mass of rock strata into any sensible form. It was the early nineteenth-century 'civil engineer and mineral surveyor'—as he styled himself—William Smith (1769–1839) who recognized that successive levels of strata are characterized by different fossils. Smith used this information to help construct the world's first geological map—of Great Britain as it happened—tracing great rock units across the country from coast to coast, each unit having a 'fingerprint' of distinctive fossils. Smith's discovery of the logic and order of fossils within strata has since been developed into a sophisticated science, and this guide to the fourth dimension has become—among other things—the glue that holds ancient climate reconstructions together.

Much good it did Smith. One result of his Herculean labours in single-handedly working out the geology of an entire country (together, admittedly, with a failed business deal) was a large financial debt and a flock of impatient creditors. He was forced to sell his collection of over 3,000 fossils, carefully arranged by stratal level, to the nation. For this he obtained £500 in total—quite a sum of money in those days, but not nearly enough. Disaster was staved off for a while—but the next year he was thrown into a debtors' prison anyway.[1]

The knowledge that different fossils characterize rocks of different ages now forms the bedrock of the geological timescale. That

Phanerozoic timescale is subdivided into intervals (some, now, only a fraction of a million years long) each largely characterized by particular types of fossil animals and plants—William Smith was right. What we know now (and William Smith did not) is that nearly all the major turnovers in the geological record, the extinctions and originations that define the boundaries between geological periods such as the Ordovician and Silurian, are, in some way, intimately related to climate change.

And what would have fascinated William Smith, for sure, is that the fossils he collected with such labour and such care were sensitive to a kind of tectonic weather that seems, in some way to have governed the pattern of the world's climate since (at least) the dawn of the Phanerozoic. The fossils felt this weather in their bones—indeed, it could change what their bones were made of. It is all down to the calcite metronome, which appears to have beaten out the slowest pulse in global climate.

The calcite metronome

There is a mineral metronome that is the most unhurried of timekeepers. It swings back and forth so slowly that its rhythm is quite imperceptible against human lifespans. For all its sluggishness, though, it seems to be an integral part of the complex global machinery that paces changes in the chemistry of the oceans and atmosphere, to tick off the transitions between icehouse and greenhouse worlds. These changes were sensed and blindly logged by the animals that lived in the seas long ago.

The oceans are saturated with positive ions of calcium and with negative ions of bicarbonate (HCO_3^-). Marine snails, molluscs such as scallops, and sea urchins, to name just a few, make their skeletons of calcium carbonate from these ions: these are the *calcifying* organisms. Later still in the geological record, calcifying marine plankton such as foraminifera would come to dominate the surface of the oceans as well as the floors of warm shallow seas, but this step was not taken until after the Palaeozoic Era. Calcium carbonate can also form in the oceans by inorganic processes and will precipitate easily in warm waters.

The calcium carbonate skeleton is one of two mineral forms, aragonite or calcite. Aragonite and calcite are polymorphs—they have exactly the same chemical composition, but they have different crystal shapes and different properties. Aragonite dissolves more easily in water than calcite. Once it is buried within rock strata, it will sooner or later dissolve, and calcite often crystallizes to take its place. For this reason aragonite is not commonly found in Mesozoic rocks, and scarcely at all in the Palaeozoic. We can still discern its presence in such ancient strata, though, because it leaves a ghost of itself, in the form of shell-shapes in the rock that are hollow or are filled with coarse secondary crystals of calcite.

Observation of limestones formed over the past 600 million years shows that seas have sometimes been dominantly aragonite-producing and at other times mainly calcite-producing. This influenced the compositions of the shells that developed, often into lasting patterns. Organisms that evolved when aragonite was dominant continued to produce aragonite shells even when the seas reverted to calcite and vice versa. Trilobites, those fantastical woodlouse-shaped arthropods that appeared in the Cambrian seas 525 million years ago were still making calcite skeletons 270 million years later in the Permian, just before they became extinct. The tiny planktonic foraminifera that evolved in the calcite seas of the Mesozoic 160 million years ago still produce calcite skeletons today, although the seas are now aragonite-dominated.

Some types of sedimentary grain, though, recorded the change quite faithfully. Ooids (sometimes called ooliths) are the sand grains that you would have between your toes on a holiday in the Bahamas. They are millimetre-sized finely layered spheres of calcium carbonate that can grow in their billions in warm shallow seas, precipitating from the sea water a little like hailstones form in the air. Those forming today are all of aragonite, while those that formed, say, in the warm Jurassic Period, when dinosaurs walked the Earth, are all of calcite. The differences between aragonite and calcite ooids in ancient limestones are clearly visible to a geologist armed with a microscope. Such particles betray the chemistry of ancient oceans even more clearly than do fossil shells.

The crystallization patterns of marine minerals do not affect the Earth's climate directly—but they are a reflection of an underlying mechanism that can push the Earth towards greater warmth or cold, or increased humidity or aridity. This mechanism is the engine of the Earth itself.

As the tectonic plates wander across the surface of the Earth there have been long periods (tens to hundreds of millions of years) when the oceans have been more vigorously building oceanic crust, tearing large continental masses apart. And, there have been other intervals of time when the continents were undergoing phases of collision, destroying the oceanic crust between them and raising great mountain chains like the ancient Appalachians or the modern Himalayas. The balance between construction and destruction tilts first to one side, then to the other.

When the ocean plates are forming, new basaltic volcanic rock is produced at the mid-ocean ridges. At the ridges, which are essentially newly formed volcanic edifices, sea water is absorbed and cycled through the fresh, still-hot basaltic rock masses. These rocks absorb the magnesium that occurs naturally in sea water. This favours a calcite sea, for calcite cannot tolerate high levels of magnesium, and forms more easily when there is less of this element in the sea water. It follows that when the oceanic crust is forming at a rapid pace, the seas tend to be poorer in magnesium, and the oceans favour the formation of calcite.

When magnesium levels are high in the sea, this element becomes more common in the atomic lattice of calcite, until calcite stops crystallizing. Now aragonite begins to form, its atomic structure not being threatened by magnesium. When oceanic crust is being more actively destroyed, and continents collide, the seas tend to be magnesium-rich as the mid-ocean ridges take up less of this element. The oceans are then more favourable to aragonite.

The tectonic patterns are linked to climate.[2] At quickly spreading mid-ocean ridges, the enhanced volcanism vents carbon dioxide at increased rates into the oceans. The carbon dioxide leaks into the atmosphere where its levels build up, and the climate moves to a more greenhouse-like state. Intervals of calcite seas, then, are associated with more quickly growing ocean crust and with greenhouse climates. Conversely, when oceanic crust is destroyed, less carbon dioxide is produced from the mid-ocean ridges. Moreover, as mountain chains rise, their silicate minerals—on weathering—absorb carbon dioxide from the air, which cools the climate. Aragonite seas therefore tend to be associated with globally cool periods often characterized by mountain-building and an icehouse climate.

More than 600 million years of ocean history can be read from the Earth's limestone strata. For much of the late Proterozoic, the seas were aragonite-forming. Then, for most of the early Palaeozoic, when the world was mainly warm (with the exception of the relatively short-lived icehouse interval around the boundary of the Ordovician and Silurian periods), they were calcite-forming. The late Palaeozoic seas were aragonite-forming again, coinciding with the longest-lived glaciation of the last half-billion years. Then, the greenhouse world of the Mesozoic saw calcite seas. And finally, in the present Cenozoic Era (that we live in today), when mountain belts have grown apace, from the Alps to the Himalayas and the Sierra Nevada to Patagonia, the oceans became aragonite-forming once more. This state persists today—though, as we shall see in Chapter 10, it might now be under threat, as humanity alters the world's chemistry.

The oscillation between calcite and aragonite seas in Earth history arose out of the changing balance between ocean crust formation and destruction. The plate tectonic motions that created these patterns also created the changing faces of the Earth, as the oceans and continents continually altered their form and pattern.

A different world

The geography of the early Palaeozoic Earth was quite different from ours. If we were able to look on to the face of that ancient Earth from space, 1,000 kilometres up and half a billion years back, there is virtually nothing that we would recognize. True, quite a lot of the basic substructures of our modern continents had formed. But, they were all in foreign parts, and engaged in unexpected alliances. The ancestors of our landmasses were mostly present, but elsewhere. They were dominated, then, by a giant called Gondwana.

Looking down from the heights of space into the depths of time, we would see Gondwana stretching from the geographic South Pole almost to the equator. Big as it was, it was just a fragment—the largest remaining one—of the once-mighty Rodinia supercontinent of late Proterozoic times. Nevertheless, Gondwana was still colossal, its various parts formed of what we now call Africa, South America, India, Saudi Arabia, Australia, and Antarctica. Gondwana, 'the forests of the Gonds' (of the Gondi people of central India, that is), was originally recognized by the influential Austrian geologist Eduard Suess (1831–1914). Suess lived and worked in Vienna, the heart of Europe, and adapted himself well to that cultural crossroads; reputedly, as a young man, he learnt one new language every year. He saw the world as a whole, and recognized that it had undergone major changes—for example that the rocks that made up the Alps had once lain at the bottom of a wide ocean, that he named the Tethys Ocean, of which the present-day Mediterranean is but a remnant.

Suess recognized Gondwana because its fossils—now widely scattered on the far-drifted fragments of this former giant landmass—

formed a distinctive and coherent grouping. Working even before Wegener's time, he did not consider that Gondwana had split and its parts separated: rather, he thought that the intervening parts, now ocean, had been land once and had simply sunk below the waves.

To the north of Gondwana lay a cluster of smaller continents that straddled the ancient equator. In the west of the tropics lay Laurentia, the ancient core of modern North America. To the east there was Baltica—the nucleus of Scandinavia and northern Europe, and Siberia. Still further east were the separated continents of North and South China. A group of yet smaller continents was scattered between these, among them a continental sliver called Avalonia, the ancient foundation of southern Ireland, Wales, and England. To the far north, and beyond the horizon, was a water-world of ocean, for no substantial landmass lay in the northern polar region of this world.

That pattern of continents and oceans helped control the climate of the times, by the way they interacted with the currents of air and water.[3] Within the early Palaeozoic oceans these currents were driven by the same forces that control them today: the heat of the Sun and the spin of the Earth. At the equator the sea and air moved from east to west as the Earth spun on its axis. To the north of the equator those winds and seas would move with a clockwise motion, and those south of the equator would adopt an anticlockwise motion. The continents would disrupt this simple pattern, deflecting warm surface water currents further south, or cold surface water currents further north.

As the warm tropical surface waters of these ancient seas travelled north and south, evaporation at their surface made them become cooler and saltier—and so more dense—and they began to sink below the less dense polar waters. This was the start of a *thermohaline* circulation (the word means heat + salt). This mechanism operates strongly when the poles are cold but is weak when

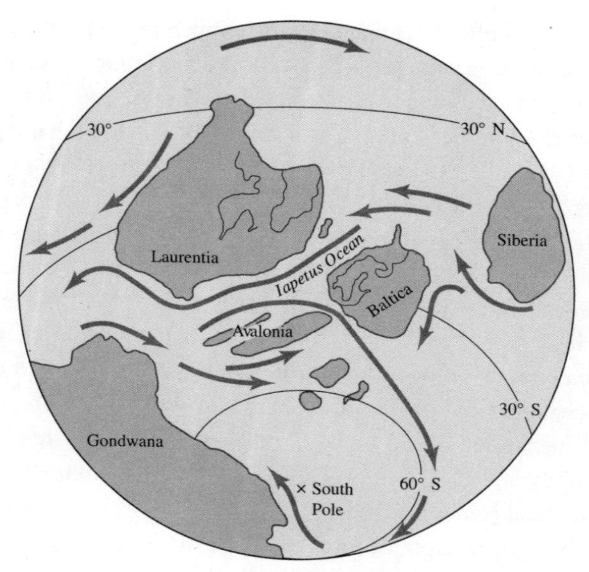

FIG. 9. The continents and oceans of the late Ordovician world. Ocean currents are indicated by the arrows. Baltica is modern Scandinavia, whilst Avalonia represents various parts of England, Wales, and Ireland. Scotland lies on the margin of Laurentia (modern North America).

the poles are warm, as the density contrast between tropical and polar waters then becomes much smaller. So, in a greenhouse climate there is weak thermohaline circulation and the oceans are sluggish and can become stagnant in places. By contrast, in an icehouse climate like ours, this circulation is strong and the oceans are well mixed.

Between greenhouse and icehouse intervals, too, the tropical ocean surface stays warm and may fluctuate by just a few degrees. But at higher latitudes, far to the north and south, the ocean surface warms substantially in a greenhouse climate, and the gradient of temperature change from equator to pole becomes shallow. The poles in a greenhouse climate may be temperate, with balmy summers and ice-free seas. But in an icehouse world the polar regions are cold, and icy waters may extend into mid-latitudes.

After the intense glaciations of Snowball Earth, global climate seems to have been generally warm. Continents were fragmenting and the seas were of calcite-forming type. The rapid growth of ocean crust may have released large volumes of volcanic carbon dioxide to the atmosphere—though this is currently just surmise. There are no reliable estimates of the amount of carbon dioxide in the atmosphere back then. But the situation was probably not simple, as the rock succession, going into and through the Cambrian, shows sea levels changing and phases of extinction as well as radiation of the world's new animals. So while Cambrian climate may have been dominantly warm, it was probably far from stable and there are indications of colder interludes.[4]

We are simply short of tools to analyse the Cambrian climate. Even though a wave of new life forms had just colonized the Earth, these did not—at first—include a diverse (and fossilizable) plankton. There were the acritarchs—planktonic single-celled algae that developed resistant outer casings and were possible forerunners of the modern dinoflagellates (the kind of organisms that today can cause toxic 'red tides'). But the acritarchs do not—yet—give us many clues about the oceans they lived in.

The succeeding Ordovician Period, though, that started about 488 million years ago, showed a change from early warmth to the first major glaciation of the Phanerozoic at its end, 40 million years later, when ice left thick tillites and ice-scratched pavements on south America and southern Africa. At the beginning of the Ordovician, too, a new type of plankton arose, and these proved more eloquent about their marine home, and how it reacted to the growing ice.

These are the graptolites—Earth's first macro-zooplankton, and rather bizarre ones at that. Originally dwellers of the seabed, some made the leap into the plankton at the beginning of the Ordovician. They thrived, and diversified. From a few early forms there soon arose hundreds of species. These swarmed in the surface of the oceans from the tropics to the polar regions, each one a tiny record of the water in which it lived.

Graptolites and climate belts

Within rock strata, fossilized graptolites look like miniature flattened fretsaw-blades. Graptolites lived together in colonies a few centimetres long, building many individual chambers (i.e. the fretsaw 'teeth') of their sturdy collective 'home'. The tiny animals that did the building (for these were sophisticated animal architects[5]) are almost never preserved, but it's likely that they looked similar to the soft parts of a tiny coral, though they are more closely related to an obscure present-day marine group, the pterobranchs. Within the colony they were probably connected, like a team of mountain climbers that are roped together. Graptolites appeared in a huge range of designs, and these designs evolved rapidly as the graptolites adapted to the shifting demands of their ocean habitat.

They spread widely across the oceans of the Ordovician and Silurian periods,[6] dying out midway through the subsequent Devonian Period, some 405 million years ago. Many of them lived in the upper part of the ocean, which was stirred by the winds and warmed by the sun.

Today, patterns of plankton distribution at the surface of modern seas reflect ocean temperature. Some plankton is adapted to the cool surface waters of the polar regions, others to the mid-latitudes, and still others to the tropics. These plankton biotopes move northwards and southwards as the climate belts narrow or expand with the waxing and waning of ice sheets. The boundaries between both the water masses and the biotopes they contain may be sharp at

Fig. 10. Silurian graptolites on a rock slab from central Wales.

ocean fronts, where water masses of different properties meet. Such patterns have existed as long as there have been oceans, but they are first clearly detected in the Ordovician Period, with the help of the graptolites.

The task, though, was not easily done. It involved comparing all the records of graptolites of that time period from around the world with each other—like an enormous, multidimensional jigsaw where not all the pieces fitted, and where one didn't have any idea of the picture that would result. Nevertheless, a young researcher from Belgium, Thijs Vandenbroucke, collaborating with Howard Armstrong, a geologist at Durham, worked on the puzzle pieces of graptolite data that almost two centuries of previous study had bequeathed, employed some ingenious statistical methods, and (perhaps a little to his surprise) a consistent picture emerged.[7]

Early in the history of plankton colonization, graptolites formed two broad biotopes, one in the surface waters of the oceans at low latitudes and the other at high latitudes. This suggests that the early Ordovician oceans were warm at the surface, with many graptolites extending over wide areas. But this pattern did not persist, and by 460 million years ago a clear change took place. Graptolites show a steep temperature gradient for the surface of the Ordovician oceans—one that looks very similar to that of the present day. And from then until near the end of the Ordovician 443 million years ago, climate cooled dramatically. The graptolites had responded to the entry of the Early Palaeozoic Icehouse.

To get another view of how this icehouse evolved, though, one does not need to analyse thousands of palaeontological records. A simple if longish hike across the countryside will do. Providing, that is, that you are in the right part of the world.

A Welsh journey

Palaeozoic geology began in Wales. All three geological periods of the early Palaeozoic are Welsh: the Cambrian is the old Roman name for Wales; and the Ordovician and Silurian are named after ancient Welsh tribes. The rocks of ancient Wales formed in the seas off Avalonia, a small continent that broke away from Gondwana during the early Ordovician, gradually drifting northwards across the expanse of the Iapetus Ocean. By the late Ordovician, Avalonia was in mid-southern latitudes, remote from the ice sheets that were growing on Gondwana. But even in Avalonia, thousands of kilometres from the source of severe cold, the influence of ice was clearly signalled.

To read this climate narrative in the rocks of Wales, we have to go north from the small coastal town of Newport, Pembrokeshire, through Cardigan, to the tiny coastal hamlet of Llangranog some 60 kilometres to the north as the crow flies. The rocks exposed along the magnificent Pembrokeshire and Cardiganshire coastline are folded and contorted now, a result of one of the collisions that Avalonia suffered on its journey. But we know from the fossils that they contain—using the principles of biostratigraphy established by William Smith—that the oldest rocks are in the south, and that the rocks gradually get younger northwards: perhaps twenty million years of geological time separates Newport from Llangranog. Today, it's a few days' hike—or, you can jump into a car and drive that distance in an hour.

The rocks of this Welsh coastline are complex and distorted. A century and a half ago, the first maps were made by geologists travelling in horse and cart, and many of the details escaped them. The rocks were shown as one great mass of slate, coloured as a block of purple. In the last few decades, though, the scientists of the British Geological Survey have brought these rocks to life, building their maps of the geology from the meagre inland rock exposures and tracking these back to the magnificent exposures of the cliffs (that, incidentally, generally have to be reached by boat). Two geologists were at the heart of this work.

If you had visited this coast just ten years ago you might have seen the two of them, a silver-haired Dick Waters and a youthful Jerry Davies hammering the time-hardened mudrock strata to look for subtle differences in texture, colour, and grain size. Both men could be seen striding across the hills together, map board in hand, geological hammers ready at their sides, undeterred by wind or rain. Dick and Jerry stood on the shoulders of their geological forebears and saw quite some way further. They could look at a mangled piece of Ordovician coastline at Llangranog, and within minutes would relate the rocks to every other bit of geology of that age across Wales, building up a picture of the events as they had unfolded hundreds of millions of years ago.[8] Such expertise comes only from a lifetime of studying the rocks of one region.

necessity of scaling up?

Between Newport and Cardigan, the coastal cliffs are, quite literally, darkly mysterious. The contorted mudrocks they display formed as carbon-rich muds in the deep sea. Seen in the cliffs, they seem black and featureless. Tap them gently with a hammer, though, and they will reveal traces of life. The rock surfaces, here and there, teem with the tell-tale fretsaw-blade-like shapes of the graptolites. The seabed on which these ancient muds accumulated was barren. There are no signs of burrowing, scratching, or scraping made by any seabed-living creatures. The graptolites, originally living in the waters above the seabed, are preserved because there was nothing at the sea bottom to eat them: no oxygen-rich surface water was reaching the bottom. These dark rocks speak of a sea where oxygen did not travel easily to the seabed. Such seas typify greenhouse worlds where there is only a sluggish thermohaline circulation of ocean currents.

Make your way north from Cardigan to the beach at Tresaith. The coast becomes a paler grey. Something fundamental has changed. The grey rocks here lie physically above the black, graptolite-rich mudrocks of Cardigan. They are full of dark streaks—ancient burrows of some unknown sea-floor animal; graptolites, though, become exceedingly rare, only here and there being represented by a few scraps. The sudden change from dark to grey, from barren to burrowed, signals oxygen on the seabed. Water from surface levels now penetrated the deep and reached the seabed where these grey rocks were deposited. The sea floor became colonized by animals that looked on the graptolites falling from the surface waters above as a ready meal—much to the frustration of the palaeontologist searching for their remains almost half a billion years later.

The grey rocks north of Cardigan tell of a world where oxygen began to be delivered to the deep sea. The most effective mechanism for this is a speeding-up of thermohaline circulation. That in turn typically signals the spread of sea ice at high latitudes; on freezing, this releases cold, dense, salty water that then sinks to the sea floor.

FIG. 11. Slumped sedimentary rocks at the north end of the bay at Llangranog, western Wales. The ancient sediments that form these rocks were crumpled as they slid down underwater slopes of Avalonia in late Ordovician times, destabilized by the falling sea level as the distant Gondwanan ice sheets grew.

It is the start of a pump that can spread oxygen-rich waters from the surface through to the ocean depths.

Further to the north still, the cliff drops down to the seaside village of Llangranog and the cliffs on the north side of the tiny bay now show a mass of intensely folded and crumpled strata. It is as though the rock gave up trying to hold itself up in layers, and simply collapsed under its own weight. That, it seems, was almost exactly what happened. Some 443 million years ago, the Gondwanan ice sheet, forming thousands of kilometres away, was reaching its peak, and its effects were transmitted literally to the other side of the world, to far warmer regions, to crumple those just-formed strata as effectively as if they had been squeezed in a vice.

And so they have—though the sediments themselves were deformed by the vice of their own weight. Here, as ice grew massively on distant Gondwana, water was taken from the sea. The sea level dropped, rapidly, by 100 metres or more. Large areas of what had been a submarine slope were exposed, and the sediments on them, no longer supported by water, became in effect much heavier. Soft, waterlogged, and unsupported, they slid down the slope, distorting and crumpling on the way. It is a signal widely seen at this level across Wales (and indeed, across other places in the world that possessed such a steep marine geography).

At low tide you may walk around the cliff on the northern headland at Llangranog. The rocks change back to black, and they are filled with graptolites again. The ancient sea has flooded back in, the ice sheet has decayed and the melt water has flowed back into the oceans. But the great diversity of Ordovician graptolites has gone. Only one or two species survived. The graptolites, and many other marine creatures, were decimated by the brief, fierce climate oscillation. The extinction event worldwide was catastrophic, wiping out more than three-quarters of the species on the planet, and it was the defining final moment of the Ordovician.

The Gondwanan ice sheet had not gone completely. It would persist perhaps another 20 million years, waxing and waning like the great ice sheets of the Cenozoic, although none of the ice advances to follow matched that at the end of the Ordovician in speed or scale. This changing ice cover at the South Pole is signalled in the Silurian rocks, that are dark and graptolitic at some levels, when sea levels were high and thermohaline circulation sluggish, and at other levels grey and burrowed, as distant ice grew and thermohaline circulation speeded up.

The glaciations of this Early Palaeozoic Icehouse may be tracked also by another signal. It is quite invisible to the naked eye, and resides in the chemistry of the rocks. It lies within carbon.

Carbon

The signal of the Ordovician glaciation, and indeed of other types of climate change, can be tracked from carbon. Carbon has three naturally occurring isotopes, the unstable and radioactive ^{14}C, beloved of 'dating archaeologists', and the stable ^{13}C and ^{12}C. The lighter isotope (^{12}C) is easier for plants to use in photosynthesis: it literally takes less energy to pick up this isotope from the surrounding air or sea water and convert it into biomass than the heavier form, so plants both small and large always possess a light carbon isotope signal, of more ^{12}C relative to ^{13}C. This is important, because changes in the isotopic ratio of ^{13}C to ^{12}C preserved in ancient rocks can be used to say something about ancient photosynthetic bio-productivity, about the way in which carbon is buried in sediments and thus removed from the oceans and air—and about the way in which carbon is recycled back into the sea water or air if sedimentary deposits are then subsequently eroded or weathered.

In the oceans, it is in the sunlit surface waters where plants—microscopic planktonic algae—take in carbon through photosynthesis. These surface waters, thus, become gradually depleted in light carbon, as this isotope is taken in by the plankton. On dying, the plankton sink into deep water: their organic skeletons dissolve, and the carbon enters the water. By this process the deep oceans become enriched in ^{12}C, while the surface ocean is depleted. This process can be measured in the modern oceans and it is called the 'biological pump'.

Modern oceans, though, are well stirred, and the carbon in the surface and deep waters is constantly being mixed, so that the gradient from top to bottom is never great. But in an ocean where the thermohaline circulation is restricted, in a greenhouse world, the surface of the oceans may develop a large carbon isotope difference to those waters of the deep ocean. Changes in ocean state, thus, may be reflected in the ratio of ^{12}C to ^{13}C of rocks deposited at the time.

In the Early Palaeozoic Icehouse, peak glacial intervals are marked by strong changes in carbon isotopes in the rocks—they become markedly enriched in the heavy carbon isotope, ^{13}C.[9] This characteristic signature is currently thought to be an effect of the major falls in sea level that took place as phases of ice growth took place. This converted what had been shallow marine areas into exposed land surfaces. Many of these areas were of limestone rich in the heavy carbon isotope, partly because they formed in warm waters that became enriched in ^{13}C as the carbon pump was carrying the light ^{12}C isotope into deep waters. On being weathered, the exposed limestones released their freight of heavy carbon back into the sea—carbon that then entered the sediments on the sea floor.

It is a complex mechanism, certainly—though also a fair reflection of a complex Earth. The signal, though, seems consistent. Marine rock strata of the early Palaeozoic show a zigzag pattern of positive carbon isotope excursions which typically signal advances of ice at high latitudes. It is a *practical* means—at least in these days of advanced technology—of tracking the vagaries of ancient climate. And it provides one more clue to puzzle over, as we try to work out just *why* climate changed as it did.

Causes

The course of the Early Palaeozoic Icehouse is becoming increasingly clear as the strata are ever more closely scrutinized. What, though, caused this climate change? To be quite blunt, we do not know (the 'why' questions in geology are always so much harder than the 'what happened' questions, especially at such a distant remove in time). Some possible scenarios may be sketched, though, from the evidence that remains.

One scenario suggests that the cooling during the late Ordovician was caused by increased weathering of silicate minerals on land, drawing down carbon dioxide from the atmosphere. Although we have few means of direct measurements of Ordovician atmospheric carbon dioxide, we can deduce that weathering of rock was increasing into the late Ordovician. There are chemical clues, locked in the strata that formed in those times that hint that such a process was then taking place. The evidence lies in another pattern of isotopes.

This time it is the metal strontium at centre stage. Strontium is an element that, if not exactly possessing magical powers, at least has a magical side to its ancestry. It was discovered by the chemist Thomas Charles Hope in 1792, who named it after Strontian, a village in the west of Scotland, where its ore (strontianite) can be found. Strontian in Gaelic is *Sron an t-Sithein*, meaning the hill of the *aos sidhe*, a mythological race of elf-like beings that live in a parallel world underground. Representing the spirits of nature, these are 'the Fair Folk'—though care should be taken not to anger them. It seems appropriate that their element be used to gauge the various moods of Mother Earth.

Two of its isotopes of strontium, the heavier ^{87}Sr and the lighter ^{86}Sr, can leave a signal preserved in ancient marine strata. The heavier isotope is formed from the radioactive decay of ^{87}Rb (an isotope of the metal rubidium) that is found in continental rocks. The lighter isotope (^{86}Sr) is yielded from hot fluids emanating from ocean crust. Therefore, by comparing the ratio of the two strontium isotopes, one can get a sense of how much continental weathering was taking place, relative to the input from the oceans.

The strontium signal in Ordovician strata shows a rising curve of greater oceanic input for much of the Ordovician, which can be said to reflect the growth of large oceans such as Iapetus through much of this time. But there is a change in this trend about 460 million years ago, when the proportion of continentally sourced strontium begins to rise. The rise is small at first, but suddenly increases 443 million years ago.[10] This increase may coincide with the time when the ancient continents of Avalonia, Baltica, and Laurentia collided and formed the Caledonian and Appalachian mountains. These mountains can still be seen in Scotland, Norway and along the east coast of the United States. Now reduced to stumps, but once the equal of the Alps, these mountains shed their ^{87}Sr into the oceans. And with that, the associated weathering of silicate minerals from the uplifted mountains may have scrubbed sufficient carbon dioxide from the atmosphere to lower it to a threshold, a tipping point where the ice could form and the polar front could shift into lower latitudes.

This is a plausible mechanism—but as we will see later (Ch. 5), it is also one that, examined more closely, is neither simple nor inevitable. However, the increased amount of weathered debris pouring into the ocean during the late Ordovician may have triggered a further effect. The fine particles of mineral matter are a nutrient, and this can stimulate the increased growth of marine phytoplankton. As these bloom, they take more carbon out of the atmosphere to build their tiny organic skeletons. When these fall to the sea floor and are buried, carbon is removed from the surface system.

Indeed, the alternating pattern of deposition of the dark (carbon-rich) and pale (carbon-poor) mudrocks in the early Palaeozoic suggests a natural feedback mechanism may have been at work, successively raising and then lowering carbon dioxide levels to act as a natural thermostat. This was a concept developed by a perceptive young Ph.D. student, Alex Page, working at the University of Leicester. A keen sportsman as well as scientist, he noted that these episodes of carbon burial were worldwide. As the black shales continued to form in the warm, sluggish seas, entombing both carbon and graptolites, Alex suggested that carbon dioxide would be drawn from the atmosphere—to the point at which glaciation started. As the ice grew on land and as sea-ice at high latitudes, ocean circulation would speed up, the sea floor would be more widely oxygenated, thus ceasing to be so efficient at burying carbon (because more of the organic carbon would be oxidized at the sea floor, released back into the ocean waters, and from there pass into the atmosphere). Atmospheric carbon dioxide levels would climb, raising temperatures and causing the ice to recede.[11] This idea of a natural global thermostat has probably more than a germ of truth in it. Based as it is on the record of the rock strata, it can be tested (to destruction, if that is where the evidence leads) by future research.

For now, these are simply plausible scenarios. It is time to move on, within a world growing ever more familiar. The carbon cycle is about to step onto dry land, allowing cause and effect in global climate to be assessed with a little less uncertainty.

The green land

For over four billion years, since the continents first formed, the land was, essentially, barren. For eons, dangerous ultraviolet light prevented the development of any life on the land. Only after the atmosphere became oxygenated, and ozone developed, was there a possibility of terrestrial life. Late in the Proterozoic, simple land-based ecosystems may have clung close to water, but the land remained largely uninhabited by plant or animal. In the early Palaeozoic, this began to change. One of the great colonization events in Earth history was about to take place—the greening of the land (and, almost immediately after, its infestation by insects and millipedes, and then by vertebrates, hauling themselves rather more slowly out of the sea).

There is evidence for life getting onto the land from the late Proterozoic. Tiny organic microfossils preserved in marine rocks of this age have been interpreted as derived from land plants. Small fossilized track-ways appear in the Ordovician, which suggest something then had gained a toehold on land. Perhaps this colonization began in water, with small inland seas that, once isolated from the ocean, freshened over time, forcing the organisms living in those lakes to adapt to fresh water or die. Or perhaps animals and plants were taking the initiative, gradually moving up rivers and then spreading on to the adjacent land.

Whatever route was taken, a galaxy of miniature forests accompanied by early millipedes and insects had spread, by the Silurian, onto the coastal areas of the land. At first, the biomass of carbon locked into these early land plants was small. Then, the tiny moss forests gave way to larger forests of ferns. By late Devonian times, some 380 million years ago, plants had developed strong and deep rooting systems. By the Carboniferous, ferns had evolved to become gigantic. The carbon sequestered in their skeletons, and then buried as coal, would become a new force in climate history.[12]

For some time, the later Silurian and perhaps much of the Devonian remained globally warm. But the Earth began to return to an icehouse in the late Devonian, and that heralded the main event, the longest-lived glaciation of the last half-billion years, that came to occupy much of late Palaeozoic time.

The ice returned on a supercontinent reborn, as continents earlier split by plate tectonics were reassembled. As the Late Palaeozoic Icehouse began to grip the Earth, continents straddled the globe from the northern to southern polar regions. In the south, there still lay the vast continent of Gondwana: this was not to meet its final demise until the Mesozoic Era (see Ch. 4). To the north, and lying across the equator, was Laurussia, formed from the assembly of ancient Laurentia, Baltica, and Avalonia that collided in the late Ordovician and Silurian. Further north was Siberia, still a separate continent and reaching almost to the North Pole.

This world was to change fundamentally through Carboniferous and Permian time. Firstly, Gondwana would collide with Laurussia forming the Hercynian Mountains whose core can still be seen stretching from south-west England to the Montagne Noire of France and on to the Harz and Sudetan mountains of central Europe. Still later, Siberia would join with Laurussia, forming a long mountain chain we now call the Urals. In the Permian world there would be one vast continent again—Rodinia reborn but renamed Pangaea. These plate tectonic changes played their part in forcing a change in Earth's climate.

The Late Palaeozoic Icehouse

The Late Palaeozoic Icehouse lasted a very long time. Ice waxed and waned across the polar regions from the late Devonian over 360 million years ago, until the mid-Permian, 100 million years later. At first the extent of the ice was small. Then, around 325 million years ago the ice sheets suddenly expanded to cover large areas of the southern Gondwana continent, and for the next 35 million years that ice persisted. Leaving behind its characteristic debris of boulder clays and gravels, it expanded still further about 300 million years ago, forming on the northern Siberia continent, and therefore making this the first truly bipolar glaciation of the Phanerozoic.

As with the Early Palaeozoic Icehouse (and the Cenozoic Icehouse to follow) the ice, once formed, was not static. Long intervals of ice sheet growth were punctuated by warm intervals when the ice sheets shrank back or perhaps disappeared altogether. These changes caused large rises and falls of sea level that are recorded around the world in the sedimentary deposits of low-lying coastal areas and shallow marine seas of the Carboniferous. When the sea level stayed the same or fell, wide vegetated landscapes spread, traversed by sluggish river and delta systems. When the sea level rose, the freshwater swamps and the forests that grew on them were drowned and then buried under sediment. The constant interplay between these trends created the coal-bearing strata that still power much of our society.

FIG. 12. Carboniferous forests: a celebrated early image from Louis Figuier's *La Terre avant la Déluge*, originally published in 1863.

What caused the late Palaeozoic glaciation? And, what forced the changes within this glaciation between intervals of cold and intervals of warmth?

There was widespread deposition of black, organic-rich mudrocks in the late Devonian and early Carboniferous, representing carbon withdrawn from the atmosphere, forcing global temperature to begin to cool. Plate tectonics probably played a role, too. Firstly, the southwards drift of some areas of Gondwana towards the pole might have enabled the initial development of ice sheets. Or, as mountains were built, increased continental weathering may have resulted (for changes in the ratio of strontium isotopes have been recorded in rocks about 325 million years old) taking more carbon dioxide from the atmosphere.

didate to bring about that deep and long-lived gla- n, remains the most obvious one. The development of rests and peat on land in low latitudes provided vast areas n storage. It was really a kind of accidental combination of gy and geography. Thus, the appearance of a substantial terrestrial flora just happened to coincide with one of the greatest tropical delta systems that ever grew on Earth, and that stretched from America in the west to China in the east, taking in Europe and Russia on the way. This provided a means to bury carbon wholesale, as the enormous coal deposits of the Carboniferous Period. Those forests, as they grew and were buried in those endless swamp muds, stripped vast quantities of carbon from the atmosphere, forcing an intensification of the late Palaeozoic Icehouse. As carbon atoms were stripped from carbon dioxide molecules and buried, oxygen levels rose in the atmosphere, probably to greater levels than today. As a result, fires were frequent in those Carboniferous forests—though wet, low-lying swampy conditions generally prevented the destruction from being catastrophic.[13]

It was this Carboniferous coal that Abraham Darby used in Coalbrookdale, Shropshire, as he founded, in 1709, the world's first industrial-scale iron-smelting factory. So began the Industrial Revolution. Later in that century, in 1789, Coalbrookdale would become the first place in the world to have an iron bridge, one that spans the River Severn. It is the great-grandfather of the Golden Gate Bridge, the Sydney Harbour Bridge, and all the other mighty steel constructions of the world. That Carboniferous coal, now powering the industrial revolutions of China and India, is the same coal that removed carbon from the late Palaeozoic atmosphere. The formation of that coal over tens of millions of years forced a major icehouse event deep in Earth's past. (Humanity's current reversal of that major turn of the carbon cycle—as that coal is extracted and converted back into carbon dioxide—is something we will turn to later in this book.)

By the late Permian the continents of Earth were reassembled into a single supercontinent, Pangaea. The interior of this supercontinent became dry, as the inland areas became, simply, too distant from sources of ocean moisture. The huge areas of humid swampland that had so influenced the Carboniferous and early Permian climate disappeared: coal now largely disappears from the fossil record until the Jurassic. The climate became warm.

Another influence on climate emerged, this time from the east. As the Permian drew to a close, the Earth's mantle, beneath the Siberian continent, released, suddenly—that is, in something under two million years—quite extraordinary amounts of basaltic lava, to cover much of that continent. The blanket of lavas became, commonly, 2–3 kilometres thick. The associated out-gassing yielded vast quantities of carbon dioxide to the atmosphere. Probably more seriously for life, there was sulphur dioxide too (that rapidly dissolved in rainwater to become sulphuric acid), hydrogen fluoride, and hydrogen chloride. Oxygen levels, too, seemed to fall then. Between these linked 'kill mechanisms' of poisoning, warming and suffocation, life on Earth underwent its greatest trial since the late Proterozoic times of Snowball Earth.[14]

The Permian–Triassic boundary is marked by the most fundamental change in fossil biota of the Phanerozoic. Greater than the calamity of the Cretaceous–Tertiary boundary, the Permian–Triassic boundary saw nine-tenths of all life on Earth extinguished. What followed next was a world with little polar ice for more than 200 million years. Earth had entered the long greenhouse of the Mesozoic.

The Last Greenhouse World

The Arctic forests

Ellesmere Island today is a destination only to a particular type of tourist: rich enough to afford the most exclusive of package tours, and hardy (or ascetic) enough to yearn for the spiritual purity of an icy wasteland, rather than the sensual pleasures of a Mediterranean seashore. The island is large—twice the size of Iceland. Yet, its largest settlement, Grise Fjord (or, in the local Inuktitut language, Aijuittuk—'the place that never thaws') has but some 140 souls—while its smallest, Eureka, bizarrely but somehow appropriately, was listed in 2006 as having precisely none.

Squeezed between northern Canada and Greenland, Ellesmere Island is well within the Arctic Circle, and its northern tip is not much more than 700 kilometres from the North Pole. A land of mountains, fjords, glaciers, and ice-fields, it has been dubbed 'the horizontal Everest'. In the short summer, the Sun never leaves the sky, and temperatures might, on brief sunny days, exceed 20 °C. When the winter months come, the Sun never rises, and temperatures drop below –40 °C. The only tree that can grow, here and there, is the dwarf Arctic willow, usually knee-high, while the mammals—musk ox, caribou, seals—have attracted Inuit hunters for some 4,000 years (and more lately, Viking explorers too).

It was the handsomely whiskered First Lieutenant Adolphus Washington Greely (1844–1935) of the United States Army who discovered the ancient forest that had lain there, deeply buried, for fifty million years, a forest as expressive of bygone glories as any Arthurian legend. As part of the First International Polar Year, in 1882, he had been given charge of a party of soldiers, and tasked with making magnetic and meteorological measurements in the far north. They explored the Greenland coast, and traversed Ellesmere Island from east to west, stumbling upon the forest in the course of these journeys. The voyage killed most of his men, and almost killed him. When the relief crews arrived, two years late (the expedition had not been ideally planned) only six men, including Greely, were left alive. They had survived—just—by eating their own boots and, it seems, the remains of their dead colleagues.

The next exploration was better organised. It was a Norwegian one, led by Otto Sverdrup (1854–1930), in the beautifully-designed little wooden ship the *Fram* (that translates as *'Forward'*), originally built to be carried within (rather than crushed by) the sea-ice for Fridtjof Nansen's classic 3-year Arctic odyssey, begun in 1893. The *Fram* came through with flying colours, and Sverdrup set sail in it in 1898 for the Canadian Arctic. He and his crew spent three years on Ellesmere Island, surviving those ferocious winters, before returning, safely and with honour, to his native Norway. The Sverdrup Islands west of Ellesmere Island are named after him, though the *sverdrup* (an oceanographic measure denoting the flow of a million cubic metres a second) was named in honour of a distant relative.

The fossils from the Paleocene forest on Ellesmere Island were hammered out of the rocks by the expedition's geologist, Per Schei. He was a heroic figure: one of the youngest of the team, his individual contribution to the geological understanding of Arctic

Canada was only exceeded once aircraft could be brought into play. His story is tragic, too—he died soon after the expedition returned, before he could write up the results. Nevertheless, those fossils, that subsequent expeditions also sought, are key to a world unlike our own. Curiously, that distant world is thrown into sharp focus because of a tectonic quirk of Ellesmere Island. While continents around have moved by perhaps a thousand kilometres, it has stayed, if not motionless, at least at very nearly the same latitude over the last fifty million years, as shown by the preserved magnetic particles in its strata.

The plant fossils are sufficiently abundant to make up substantial coal seams that the energy companies are now eyeing hungrily—to the despair of the palaeontologists wanting to preserve these sites for science. They show an Arctic forest of the Eocene that was lush, wet and green. Trees such as *Metasequoia*, the dawn redwood, reached 60 metres tall. Although a conifer, it shed its leaves in the six-month darkness of a winter where temperatures hovered around zero, its finely developed annual rings also indicating a shutdown of activity once the light disappeared. Once thought to be long extinct, it was rediscovered in 1944 in Sichuan, in China; from there, it has re-colonized the world—or at least its parks and gardens, as a popular ornamental tree.

The finely fossilized leaves betray the scale of those far northern downpours, as, on average, larger leaves form in wetter climates. Today Ellesmere Island is a true polar desert with only some 7 centimetres of precipitation annually. In Paleocene times, the evidence of the fossil leaves indicates over a metre of rainfall a year.[1] They tell stories of temperature, too, smooth-edged leaves being commoner as temperature rises: the pattern here suggests that summertime temperatures rose to some 15 °C while in the winter darkness they hovered around zero.

In the forests, roamed animals whose bones can now be found among the coal seams—alligators, giant tortoises, brontotheres, early primates, and *Coryphodon*, the last of these looking a little like a hippopotamus retouched by Walt Disney. Did they overwinter in the dark forests, or did they migrate annually, as caribou do today—south towards a little more light and warmth in the winter months? The chemistry of the bones, notably the patterns of the isotopes of carbon and oxygen, suggest that they lived through the Arctic darkness, as year-round inhabitants of the polar forests.[2]

The heavy rainfall washed off the land into the Arctic Ocean, which was then more landlocked than at present, in sufficient amounts to freshen its surface waters—and quite considerably, at that.[3] At intervals in the Eocene of the Arctic Ocean there were enormous blooms of the free-floating fern *Azolla*. It is one of the fastest-growing plants in the world, when conditions are right. The *Azolla* blooms then spread across the surface waters, their remains falling into the oozes at the bottom of the Arctic Ocean—oozes that were sampled, a few years ago, by a deep-sea drilling ship shepherded by a couple of ice-breakers.

The climate conundrum

Something like this climate pattern persisted on the Earth, more or less, throughout the whole of the Mesozoic Era and on into the early part of the Cenozoic Era. Not altogether, of course. Within this enormous span of time, of some 200 million years, it is difficult to envisage a climate remaining entirely warm and stable. There were climate variations—some quite dramatic ones, and intervals that were cooler (the later part of the Jurassic Period, for instance) when some ice seems to have grown in polar regions.[4]

Nevertheless, this was in general a greenhouse world, with the tropics a little hotter than they are today, and the high polar regions up to 40 °C warmer than now. With the absence, or near-absence, of land-based ice sheets, sea levels were of the order of 100 metres higher than today, and low-lying parts of the continents became flooded to form extensive shallow (termed 'epeiric') seas. And the oceans were warm, through and through. If one dives down in today's ocean, even at the equator, the balmy, sun-warmed waters are only at the surface. About 100 metres down it suddenly gets much cooler. The waters soon reach a frigid 4 or 5 °C, and may be only a couple of degrees above freezing in the deepest waters. The ocean today is essentially bone-chillingly cold, overlain only by a thin surface skin of sun-warmed water at low latitudes.

If you made the same dive during Mesozoic greenhouse times, the temperature difference would be far less, with deep-water temperatures ranging as high as 20 °C. The oceans of the greenhouse world were like a warm, deep bath. A calm bath, too. The strong temperature differences in today's oceans help drive a vigorous ocean circulation system that, today, takes oxygenated waters through to virtually all parts of the ocean floor (as it probably did when ice grew quickly and massively in the late Ordovician too—see Ch. 3). In the Mesozoic, the circulation systems were more sluggish and, at times, the oxygen supply to the ocean floor was effectively choked off.

More about these oxygen crises anon. But one problem with the Cretaceous world was that of its astounding geographic evenness. From equator to pole there was much less of a temperature gradient than today, and rather like the bumblebee that was (as used to be thought) not built to fly and yet somehow does, the Mesozoic world is difficult to reconstruct. Computer models of Cretaceous climate have been extremely difficult to fit with what we know of the patterns of geography, temperature, and humidity of that time.[5] The problem is basically one of heat distribution—how can the heat travel from the equator to the poles so efficiently? And of heat retention, too—how did the polar regions stay as warm as they did?

Part of the answer may lie in the enhanced greenhouse gas content. Levels of carbon dioxide in the deep past are always tricky to work out (see Chs. 5 and 6), and there are wide error bars on the methods used. But, estimates of carbon dioxide levels in the Cretaceous range from some two times higher to ten times higher than today.[6]

There is another factor. Water vapour is a powerful greenhouse gas, albeit one that fluctuates in the atmosphere (and also condenses into clouds that can both reflect incoming solar radiation and trap heat). Nevertheless, as temperatures rise, more water in general will evaporate from the seas: from seas, moreover, that have extended much further over the continents than they do today. And in at least the later parts of the Mesozoic and early Cenozoic, the high humidity and rainfall around the Arctic Circle suggest there was enhanced release of water vapour into the atmosphere—as a kind of extra blanket, perhaps, keeping the polar regions of the world warm.

The rhythms of a warm world

Grove Karl Gilbert (1843–1918) lived through the pioneering days of North American geology, in the latter part of the nineteenth century. At that time, on that continent, he grew to become perhaps the pre-eminent geologist of his time.[7] A household name?—perhaps not, and his name today would ring a bell mainly with geomorphologists of the old-fashioned type, through mention of a Gilbert-type delta, a sort of small delta typically associated with rivers flowing into lakes. And yet he ranged far wider, a scientist who lived for the outdoors, but who was fated to spend most of his life dealing (effectively, but unwillingly) with bureaucracy and administration. Indeed the reason that his name turns up in these pages is but a footnote within his life's work—a significant footnote, nevertheless, and one that reveals much about his very particular scientific qualities.

In his childhood he suffered a great deal of ill-health, and his life revolved around the family home, a refuge that he was to return to throughout his life, that his family called 'The Nutshell'. The dominant character was his father, a painter and maverick who organized games, riddles, puzzles, and poetry competitions for his children. Gilbert grew up to be studious, and analytical, and something of a mathematician and classicist. He fell into geology by accident, through being hired to help catalogue specimens at a museum, Cosmos Hall, devoted to the work of the great German scientist and explorer, Alexander von Humboldt—a man who, then, had true star quality. The work Gilbert had to do was dull, but he did it well, and was asked to carry out more ambitious tasks. He did not work, though, in the spirit of those inspired by Humboldt, glorying in the grand romantic vision of the distant past. Rather, he sought underlying patterns and processes that could be analysed logically and meticulously.

When helping to excavate a mastodon—a newly discovered monster of the ancient Earth—from a river bed near Albany, his attention turned away from the spectacular bones to the potholes in the river bed: he analysed those, to see just how, mechanically, a river could create such geometrically distinctive structures. He joined the early geological surveys of the Rockies, took part in further expeditions by foot and on muleback through the still untamed West, braving unexplored rapids and the killing heat of Death Valley, breakfasting on such as rattlesnake when food stocks were low. His researches always focused on how landscape and rock structure related to processes of erosion and deposition operating through the almost endless stretches of geological time. He analysed the Great Salt Lake and its yet larger ancestor of Ice Age times, and he examined the environmental impact of gold mining.

Time fascinated him. On his travels he recognized one key into the then uncharted depths of Earth time—marine strata of Cretaceous age in Colorado, in which he noticed that rhythmic patterns were imprinted. He deduced that the patterns were regular enough to need some kind of external pacemaker to trigger repeated changes in the kind of sediment being washed on to that ancient sea floor. He looked to the skies for the mechanism that could influence sedimentation in such a metronome-like fashion. It was the wobbling of the Earth's axis of spin, he said, that could create such a phenomenon in the strata.[8] It was the spirit of the Nutshell at work, and in making that connection he reached back to a chain of studies that is as old as science itself.

Precession of the equinoxes, that wobble is called. What does it mean, quite? Take the equinox first. It marks the changeover of the winter and summer seasons—thus, it is the point in time at which the Earth passes between the summer season at one pole and that at the other pole. Around that point in time, which happens twice a year, in late March and late September, days and nights around the world are of equal length (these points lie exactly between the solstices—the times of midsummer and midwinter). At the equinoxes, at a point on the equator, the Sun can be observed to be *exactly* overhead.

In any human lifetime, this pattern appears to be constant from year to year. We have seasons because the Earth's spin axis is not at right angles to the plane of the Earth's orbit around the Sun, but is substantially tilted, being at an angle of something more than 20 degrees to the vertical. So, from late September to March the northern hemisphere faces away from the Sun while the southern hemisphere, by contrast, basks in the Sun—only for its own winter to roll around, half a year later.

But with infinite slowness—to human perceptions, that is—the Earth's spin axis moves with respect to that gigantic frame of reference that surrounds us: the star systems of outer space. Thus, the North Pole now points towards Polaris, the Pole Star that lies at the end of the constellation known as Ursa Minor. Four thousand years ago, it pointed towards a spot between the end of Ursa Minor and that of the neighbouring constellation of Ursa Major (part of which forms the 'Big Dipper'). Two thousand years before that, it pointed towards the star at the end of the Big Dipper.

This change is not something that can be noticed by casual observation—especially these days, when most of us live in the comfortable cocoons of our artificially lit towns and cities. In the world of ancient Greece and Egypt and Babylonia, though, the stars were ever-visible at night, and loomed large in the life of the people. They were the means by which the basic rhythms of their lives, such as the timing of plantings and harvests, could be ordered.

Those technologically primitive peoples therefore developed remarkable sophistication in observing the stars and analysing their movements in the sky. It was Hipparchus, part of that great flowering of Hellenistic culture associated with the Egyptian city of Alexandria, who, sometime around 150 BC, first recognized the phenomenon of precession. We know little of him, except that for most of his life he was a mathematician and astronomer, whose groundbreaking achievements are signalled in the work of Ptolemy, three centuries later. He is generally credited with being the inventor of trigonometry, for instance.

To infer that such a thing as precession had taken place, Hipparchus needed to reach back to the work of his predecessors. He compared the astronomical measurements made by the Babylonians, centuries before, with his own measurements—and he found that they were different. Not by much—the star systems had moved by just two degrees at the time of the equinoxes. He calculated—as Ptolemy affirmed—that the movement would trace out a circle relative to the constellations in under 36,000 years.

Hipparchus was not far out. Later calculations showed that the wobble of the Earth's axis traces out a circle in the heavens that takes 26,000 years to complete. Yet to alter conditions on the Earth in such a way that would leave indelible impressions upon Gilbert's Cretaceous strata, this wobble in itself is not enough. It needs to be linked with yet another rhythmic astronomical change—an even slower movement in the Earth's orbit around the Sun. This orbit is not circular, but an ellipse, in which the Sun is not at the centre, but displaced from it by more than a million miles. The Earth is thus at different distances from the Sun at different times of the year. This ellipse, too, is moving in space, rotating very slowly—but in the opposite, anticlockwise direction. By combining these two opposing motions, the equinoxes describe a circle along the orbit once every 21,000 years.

This produces a small but significant difference—a few per cent—in the amount of sunlight received by the two hemispheres of the Earth between winter and summer. Today, the northern hemisphere winter occurs when the Earth is near the closest point of its elliptical orbit to the Sun (and thus the northern hemisphere winter is relatively warm, while that of the southern hemisphere is, of course, relatively cool). But 10,500 years ago, when the Earth's spin axis was tilted in the opposite direction relative to the orbit, it was during the northern hemisphere summers that the Earth was closest to the Sun, giving rise to slightly warmer summers—and correspondingly colder winters—on that hemisphere.

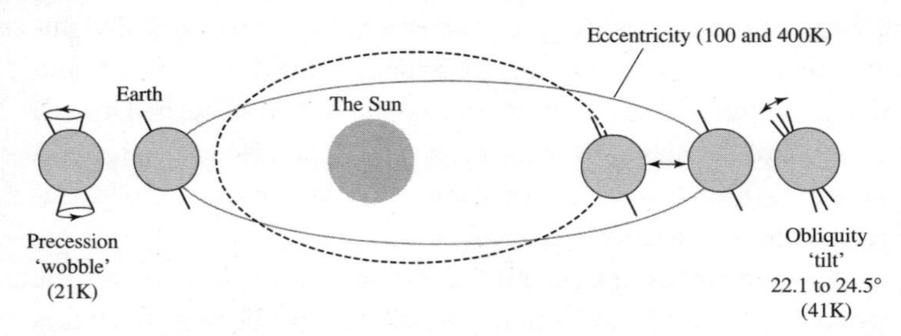

FIG. 13. Incoming heat from the Sun varies according to changes in the Earth's movement through space. Eccentricity defines the shape of the Earth's orbit around the Sun which varies from circular to elliptical. Obliquity is the tilt of the Earth's axis of rotation relative to the plane of its orbit around the Sun. Precession is the wobble of the Earth as it spins on its axis. Collectively these orbital effects modulate the amount of sunlight reaching different parts of the Earth in different seasons, and hence influence climate patterns.

These are slight changes in the seasonal distribution in sunlight between the two hemispheres and, by Gilbert's time, such effects had already been calculated by scientists such as Joseph Adhémar and Jean le Rond D'Alembert. The science was further developed by that remarkably dogged self-taught Scottish savant, James Croll (though the work of that now-famous mathematician, the Serb Milutin Milankovitch—whose name is now applied to this kind of research—was still decades in the future).

Gilbert knew of this early work, and how these scholars were already suggesting that the astronomical variations were the cause of the Pleistocene ice ages. However, the chaotic Pleistocene record of glacial boulders and sands and gravels on land was—then—too fragmentary and too poorly dated to prove anything one way or another (that was to change later, as we shall see). But, in his Cretaceous strata, over a kilometre thick, he had a long, unbroken succession showing a near-metronomic alternation of rock types. In particular, there were many, regular couplets of shale and limestone, each being a few tens of centimetres thick.

These alternations, he inferred, could best be explained by the kind of regular alternations in climate that could control the types of sediment being laid down on that sea floor. He was suitably circumspect about the nature of the climatic oscillations—changes in wind patterns, perhaps, or between wet and dry climates, or (a more fantastical speculation, this) growing ice-caps shifting the Earth's centre of gravity to alter regional sea levels. No matter—the changes were significant enough to alter conditions at the Earth's surface sufficiently to leave a clear imprint in the rock. And that climate change between two alternate states occurred *many hundreds* of times in the Cretaceous.

Gilbert's chosen mechanism to drive this climate oscillation was precession. In essence, he was correct—and far ahead of his time, for it took the best part of a century for this strand of research to be pursued further. On this basis, he suggested that the kilometre-thickness of strata that he studied, that made up a good part of the Cretaceous Period, represented some twenty-one million years. It was a bold statement, not least because this figure was very close to the latest estimate that the renowned and influential physicist William Thomson had just then published *for the entire age of the Earth*, based on how long a molten globe of rock would cool to its present temperature. Thomson's estimate then seemed unanswerably logical—but was entirely undermined a few years later when the phenomenon of radioactivity was discovered, and seen to be a heat source that could keep an Earth-sized planet molten inside for billions of years. Gilbert's calculation turned out to be remarkably close to modern estimates of the length of the late Cretaceous.

There are other types of astronomical variation. These too had been painstakingly worked out in the nineteenth century, by such means as calculating the gravitational effects of the other planets on the Earth's orbit. The amount of the Earth's tilt, for instance, varied by just a couple of degrees, from 22.1 to 24.5 degrees, and back again, every 41,000 years. And the Earth's elliptical orbit varied too, not just in its position but in its shape, stretching from near-circularity to a more pronounced ellipse and back again every 100,000 years (in the 'short eccentricity' mode) and also every 400,000 years (the 'long eccentricity mode'). Both of these additional variations give rise to slightly greater (high tilt/greater degree of eccentricity of orbit) or lesser (low tilt/more circular orbit) seasonality—and hence provide systematic, testable influences on the Earth's climate.

The fact that this mechanism is rigorously, statistically testable is one of its great strengths. This kind of research is carried out today by carefully measuring the nature and thickness of the different types of layers, and of properties that may be climatically influenced (such as the varying proportions of clay and carbonate, for instance, or the amounts of iron-bearing minerals), and then statistically analysing these. The presence of astronomical influence can be established if the sedimentary patterns show ratios of most (or preferably all) of the 400-, 100-, 41-, and 21-thousand-year rhythms.

The modern work on these patterns, by such contemporary astro-geologists as Andy Gale and Fritz Hilgen, working on (for instance) the regularly striped Cretaceous Chalk strata that can be seen in the white cliffs around Dover,[9] seem now to have established beyond doubt that these influences were ever-present controls, fine-tuning the overall warm climate of the Mesozoic and the early Cenozoic eras. The different types of sediment generally reflect local oscillations in climate, for instance between wetter and drier conditions, or to slightly warmer and colder conditions, or different wind regimes, and these in turn would influence the pattern and strength of ocean currents.

However, this greenhouse world was not always so placid. Climate could make sudden and dramatic excursions to a quite different state, before subsiding back to normal. These excursions are now becoming ever more apparent (and ever more of them are being found). They are the hyperthermals, and they show what can happen when a world that is already warm catches a fever.

The hyperthermals

The patterns of strata can have consequences in quite unexpected places—in the world of fashion, for instance. Whitby, on the Yorkshire coast, is a place with a long history. It has its own saint (Hilda, the local abbess in the seventh century) and representation from the dark side, too (Bram Stoker's Count Dracula landed there, one stormy night, in the ill-fated schooner *Demeter*, steered into the harbour by the hand of a corpse lashed to the wheel). But before any of that history—real or imagined—had taken place, the local people had noticed that one layer of shale in the cliff, 20 metres or so thick, was a little darker than the shales above and below. It was known as the Jet Rock because it contained lumps of black mineral—'jet'—that looked a little like coal, but that, unlike typical coal, could be worked into beads and other ornaments, and be polished to a black, lustrous sheen.

Jet ornaments can be found in Bronze Age burial sites, and it was popular, too, with the Romans.[10] Pliny the Elder thought it more than decorative: burning it, he declared, could drive off snakes. It reached its peak of popularity, though, in eighteenth-century England, when, in national mourning for George IV, the Lord Chamberlain declared that 'the ornament will be jet'. Queen Victoria, devastated by the loss of Prince Albert, decreed that, for a whole year, only jet jewellery could be worn in court.

The real history of jet is much older, and represents a different sort of loss. Like coal, it is a form of fossil wood, being the remains of floating lumps of driftwood (*Araucaria*, mainly—the monkey puzzle tree) that became waterlogged, then sank into a muddy sea floor, a little over 180 million years ago. This was a remarkable episode of early Jurassic history that lasted some one hundred thousand years, when many things seemed to happen more or less at the same time. There was a wave of extinctions of marine organisms. The sea warmed by some 5 °C worldwide while its level rose, perhaps by a couple of tens of metres, and much of the sea floor also became oxygen-starved.[11] Buried in the stagnant muds on a stagnant sea floor, the wood gradually altered, over millions of years, to take on its distinctive texture and sheen—to catch the eye, much later, of a distraught queen of Great Britain.

The Jet Rock is a distinctive stratum, forming a 'marker' bed that can be traced across Britain and into Europe, with equivalents being recognized further still. Although the sea floor was generally lifeless, this stratum can be recognized widely because it contains distinctive, beautiful ammonites—the spiral-shelled creatures of which the living pearly nautilus is a surviving relative. These, flourishing in life in the sunlit, oxygenated surface waters, sank into the fetid sea floor muds at depth, for that peculiar kind of immortality that is fossilization. The Whitby ammonites, so commonly found on the shore, used to be regarded as the remains of snakes that had been miraculously petrified by the formidable St Hilda—a lady who obviously shared the herpetological prejudices of Pliny the Elder—and flung from the cliff top.

They are used by geologists today as supreme time-markers, to pin the Jet Rock into the time framework of the Jurassic: to be precise, within the *Cleviceras exaratum* ammonite Subzone of the *Harpoceras serpentinum* Zone within the Toarcian Age of the early Jurassic. This may seem an unpardonably abstruse way of putting things—but to those who work on Jurassic phenomena it is a small part of the backbone of a precise timescale, which makes possible the reconstruction of a coherent history—and explanation—of events in the deep geological past.

The crucial clue to the curious mystery of the Jet Rock is invisible to the naked eye. It needs the atom-sifting capabilities of a mass spectrometer to bring it to light. The proportion of the light carbon isotope ^{12}C to heavy carbon ^{13}C rises markedly at the base of the Jet Rock to a high level somewhere near the middle of this stratum, before falling back to normal levels. This flood of light carbon into the surface environment seems to have happened all around the world at that time, both on land and at sea—for the pattern turns up both in the jet itself, that represents fragments of land plants,[12] and in the enclosing shale, in which the carbon is mostly derived from marine microplankton.

This extra carbon with its distinctive 'light' pattern must originally have been biologically derived (as living organisms preferentially absorb the light isotope of carbon from their surrounding environment), but given the calculated volume of the release (between two and four thousand billion tons) it could not have come from animals and plants then living: the biosphere, then as now, is simply not big enough. The most plausible source for it is in millions of years' worth of buried organic carbon being released, geologically suddenly, into the ocean and the atmosphere.

The finger of suspicion has been pointed at the methane bound up as waxy-textured hydrates in sea floor sediments,[13] though another potential source lies in buried coal seams and organic-rich shales. The former can be released if the sea floor warms slightly, for this destabilizes the hydrate structure, releasing methane gas, while the latter may be literally burnt by, say, a very large inrush of magma.[14] Methane is a very powerful greenhouse gas that, as we have noted previously, is soon oxidized in the atmosphere into that other greenhouse gas, carbon dioxide. Some combination of these two gases most likely gave rise to the hyper-warm conditions of Jet Rock time, and hence to the sea level rise that took place then (by simple thermal expansion of the oceans, as well as by melting any small ice-sheets that were then present). It would also have led to the slowdown of ocean circulation to cause oxygen starvation at depth. All this impacted upon the animals and plants of the day, driving some of them to extinction and changing the habitats of others. What is present in the Jet Rock is, essentially, a fossilized global warming event.

The pacing of release of the carbon can be tracked, in those Yorkshire cliffs, by very detailed analysis of the carbon patterns through the Jet Rock.[15] They did not rise smoothly, but in fits and starts that suggest the influence of the same astronomical patterns that so intrigued Grove Karl Gilbert, including the wobble of the Earth's axis that is precession. Altogether the rise to the maximum levels of injected 'light' carbon took several tens of thousands of years, though this included rapid jumps in light carbon levels that took only a few hundred years. The recovery was slower: it took something of the order of one hundred thousand years before the Earth's carbon balance, and surface temperatures and sea floor oxygenation levels, were brought back to normal.

Just how was the global climate of the Toarcian Age restored to the long-term, background state of the early Jurassic? The process, as shown by those carbon isotopes, involved slow extraction of carbon dioxide from the atmosphere and its conversion into some mineral form that could be naturally buried underground. One way of doing this is by growing plant material in the sea as plankton and burying its remains in newly formed rock strata, and this is essentially what happened in the production of the carbon-rich Jet Rock and its equivalents around the world. Thus, the rock itself and all the rich fossil material it contains is essentially a part of a planetary thermostat—a negative feedback mechanism that stabilized the Earth's climate.[16]

Another way is weathering: the neutralization of carbonic acid (dissolved carbon dioxide) in rainwater, by the reaction of this acid rain with exposed rocks at the Earth's surface. The carbon in this way is converted into dissolved carbonate and bicarbonate ions, to be washed via rivers into the sea, eventually to be precipitated as limestone strata on the sea floor. There is a strong hint that something of this sort happened in the Toarcian, because of another set of invisible and subtle isotopic clues within the Jet Rock. This involves isotopes of the element osmium, which has the distinction of being the heaviest element in the periodic table, at a staggering 22.6 grams per cubic centimetre. Its name, rather unromantically, comes from the Greek word for 'a smell'—as its discoverer, William Tennant in 1803, found its oxide form to be extremely pungent.

There is not much osmium in mudrock—only a few parts per billion—but just enough to be measured by the best atom-counting machines. There are two common isotopes—one that largely comes from ocean floor basalts, and one that comes primarily from the weathering of continental rocks. In the Jet Rock, there is a marked increase in the continent-derived isotope, which generally parallels the rise and fall of the light carbon influx. This suggests that, during the Toarcian, the rate of chemical weathering of the landscape increased markedly—because the climate was hotter and probably also wetter, and because the rain had become significantly more acid, as a result of the extra carbon dioxide in the air.[17] The osmium evidence is intriguing—but not yet cast-iron, for there are other ways of producing a pattern like this locally, such as by altering patterns of marine currents. The test will be to measure patterns of osmium isotopes in strata elsewhere in the world, to see whether they are consistent enough to point convincingly to a worldwide event of enhanced chemical weathering.

This kind of event is now termed a 'hyperthermal' because of the abrupt way in which an already warm climate can suddenly become yet hotter, with strong evidence for an outburst of greenhouse gases as the cause. And, the Toarcian event is not the only one. A very similar event took place about a hundred million years later, still within the same overall greenhouse regime, at the boundary between the Paleocene and Eocene epochs of the Cenozoic Era, almost exactly 55 million years ago.[18] Again, there is evidence of a sudden increase in both atmospheric carbon dioxide levels and global temperatures, and of significant disruptions—extinctions, migrations—to the animal and plant life both on sea and on land. One striking migration is of a marine planktonic dinoflagellate (a minute flagellum-bearing alga) called *Apectodinium*. This inhabited the warmest of water only, and in the Paleocene it was confined to the tropical belt. But, in the early Eocene it suddenly spread much further to the north and south, attaining an almost global distribution. It is therefore a clear marker for this particular moment of geological time—and a signal that the world's oceans—particularly at high latitudes—had suddenly become warmer.[19]

Because the event is so much closer to us than the Toarcian one, we see it from an additional perspective—that of the ocean floors. Our view of the Toarcian event is largely derived from shallow seas that extended over continental masses, because almost all of the ocean floors from that time have been subducted—pushed back into the Earth's mantle and destroyed, by the inexorable action of plate tectonics. But there is still plenty of Paleocene and Eocene ocean floor left in today's world. All one has to do to reach it is to drill down into the older parts of the present-day ocean floor: there, buried beneath a kilometre or so of younger oozes, are the Paleocene-Eocene

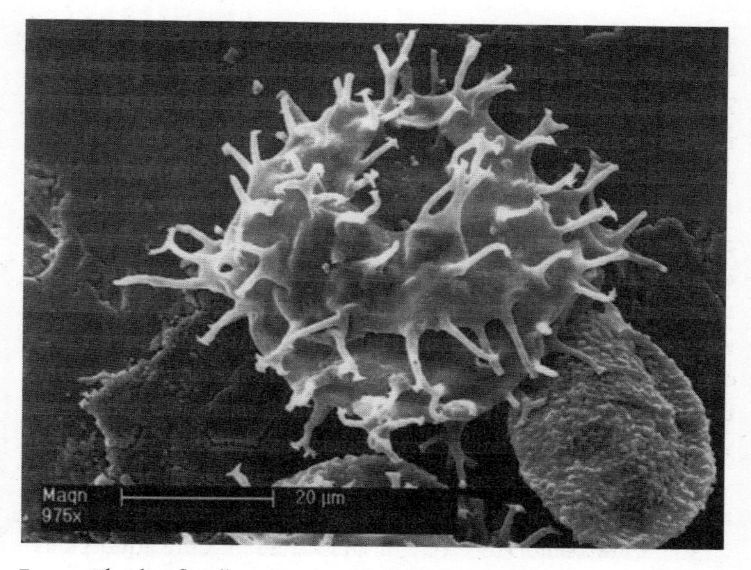

FIG. 14. The dinoflagellate cyst *Apectodinium homomorphum* from the New Jersey marine shelf. During the Paleocene–Eocene Thermal Maximum *Apectodinium* spread from the equator far to the north and south.

boundary strata. Admittedly, this is a technologically challenging and costly exercise—but its fruits, via the Ocean Drilling Programme—have been scientific gold dust. The particular set of ocean strata that mark what is now universally known to geologists as the PETM (the Paleocene-Eocene Thermal Maximum) show another of the effects of a greenhouse hyperthermal—the dissolving, literally, of large areas of sea floor by acid, corrosive bottom waters.

This phenomenon is clearly visible in borehole cores retrieved from this level.[20] The white calcareous oozes of late Paleocene age are abruptly overlain by the earliest Eocene deposits: dark muds that are made up of the finest, far-travelled clay and windblown dust that fell into the ocean, but with little or no carbonate—this was all dissolved, because the bottom waters had become more acid.

By comparing boreholes across the former ocean, one can work out the anatomy of this striking dissolution event. More specifically, it marks a sudden rise of a level within the oceans, known as the carbonate compensation depth, below which calcium carbonate is not preserved on the sea floor. Today, for instance, calcium carbonate—mainly derived from the sinking of shells of calcareous microplankton such as foraminifera—can only accumulate where the water is less deep than about 4 kilometres (a little less in the Pacific, a little more in the Atlantic). This is because the deepest ocean waters have accumulated sufficient carbon dioxide as carbonic acid (mainly from the breakdown of sinking organic matter) to become corrosive enough to dissolve the delicate foraminifer shells as they slowly sink through the water column. Above 4 kilometres, there is a kind of a 'snowline' (seen on, say, underwater seamounts or where the whole ocean floor rises towards a mid-ocean ridge), where calcareous oozes can accumulate.

Add a few trillion tons of carbon dioxide into the atmosphere, and this gas will dissolve into surface sea water, and the extra acidity will be slowly mixed into the oceans over centuries and millennia, making the entire ocean mass more acidic. This reduces the area over which calcium carbonate oozes can accumulate—in effect raising the 'carbonate snowline'. This is what can be vividly seen in the PETM hyperthermal event. As with the Toarcian of the Jet Rock, it took about a hundred thousand years for conditions of ocean temperature and acidity to return to background levels, and as with the Toarcian, it was probably some combination of the increased burial and productivity of organic matter, and rock weathering on land, that acted as a planetary thermostat. Perhaps the former mechanism was important in this event, as the duration of the PETM suggests that recovery was too rapid to be explained solely by rock weathering, and that a sudden flowering of life (and then its burial) may have helped bring temperatures down.

Other hyperthermals have been recognized, too. In the Eocene, a few smaller versions of the PETM have come to light, perhaps driven

FIG. 15. The dark rocks of a Cretaceous oceanic anoxic event, preserved within Italian chalk.

by carbon dioxide 'degassed' from the oceans.[21] And earlier, in the Cretaceous, a succession of 'oceanic anoxic events' have long been known, and have become so well studied that they have spawned their own acronym (OAEs), widely understood among geologists.[22] The Cretaceous OAEs are notable as episodes of worldwide oxygen starvation of the deep sea floor (thereby killing any multicellular organisms that inhabited those depths). These OAEs are now generally seen, also, as global warming events, accompanied by sea level rises, although their carbon isotope patterns are typically more complex than those of the Toarcian and the PETM. In some instances, they seem to follow pulses of enhanced magmatism on the ocean floors.[23]

This tendency towards geologically brief spikes of extreme global heat, associated with carbon release from ground to sea and air, and with the warming concentrated at high latitudes, seems to be a feature of the greenhouse world. Is such a tendency of significance to us now, given that the world that we live in is still essentially an icehouse world?

One similarity lies in the generally accepted causal mechanism, at least for the most pronounced events: the release of massive amounts of carbon into the atmosphere. Our human civilization has not yet emitted quite so much carbon—we have managed half a trillion tons as opposed to some two to four trillion tons calculated for the Toarcian and PETM events. But we have released this carbon much more quickly—in a mere couple of centuries as opposed to over tens of thousands of years. And, the lesson emerging from, say, the Jet Rock (when looked at in fine detail) is that the progress of these events is complex, with poorly understood feedbacks acting to both slow down and speed up the rate of climate change at different times as an event unfolds. We humans have started something that will likely go on to develop its own momentum.

Nevertheless, this long interval of some 200 million years—the whole of the Mesozoic Era and the beginning of the Cenozoic—was essentially one of global warmth. Now, it is time to move on, and see how the Earth became refrigerated. Looked at close up, it is quite a rollercoaster ride, for our world appears more complex the more closely it is examined. It is time to cross the boundary into the Oligocene Epoch, as an ice sheet grows on Antarctica.

The Ice Returns

The land of the giant penguins

Among the marvellous fossils retrieved from Seymour Island—a thin strip of land near the northern tip of the Antarctic Peninsula, is a giant penguin that lived forty million years ago. Called simply 'Nordenskiöld's giant penguin', after one of the great early Antarctic explorers, it is not the kind of animal you would like to meet down a dark alley late at night. Standing at nearly the height of an average man and with a long beak to match, it was much taller than the modern Emperor penguin. Nordenskiöld's giant penguin was a portent of a cooling climate. Its bones—many of which now reside in the collections of the Natural History Museum in London—have been found within the Eocene mudrocks of Seymour Island. This island holds a special affection for palaeoclimatologists. It was here, in the late nineteenth century, that some of the first Antarctic fossils were found.[1] These give a glimpse of what that continent was like before it became an icy wilderness.

Seventy million years ago, wide Cretaceous forests, inhabited by dinosaurs, flourished in Antarctica. Even as little as fifty million years ago, the kinds of tree and shrub that thrive today in Patagonia once covered the hills and slopes of the mountainous Antarctic Peninsula. Their fossilized remains are found in the rocks of Seymour Island. In the summer months the island is warmed by the faint Antarctic sun, its surface melting like a chocolate cake at a picnic. The resulting muddy quagmire is worth persevering with. It yields the most wonderful fossils of ancient plants, among them *Auracaria*, the warmth-loving monkey-puzzle tree. Antarctic scientists have another, ulterior motive for visiting Seymour Island; those in the know are aware that the Argentine Base at Marambio is famous for its steaks. They are the best on the continent, and everyone hopes to get invited in.

How then did Antarctica change from a continent of lush forests to a frozen wasteland? After all, this part of ancient Gondwana had already drifted over the southern polar region during the Cretaceous. Thus, Antarctica is not simply a frozen wasteland because it lies at the Pole. To understand how the land became a refrigerator, we need to enquire deeper into its history.

Discovery of the Antarctic

Until the eighteenth century little was known of a great land mass in the polar south. Ancient Greeks, sailors, and navigators had all speculated on its presence. It appeared as *Terra Australis Incognito* on some mid-eighteenth-century maps, as though preserved from some ancient folk memory. Not until the early nineteenth century was it first sighted, almost simultaneously, by three famous maritime explorers: the Russian Thaddeus von Bellingshausen, the British navy officer Edward Bransfield, and the American sealer Nathaniel Palmer. And then not until the early twentieth century did its real outline emerge, and the continent become regularly visited, at first by only those most intrepid of explorers, Roald Amundsen, Robert

Scott, and Ernest Shackleton. On one of these early voyages of discovery, the Norwegian Carl Anton Larsen discovered Seymour Island, and in 1892 his expedition collected fossils of petrified wood there.

So very little was known of the Antarctic that, even in the 1950s, Hollywood film directors could get away with making movies like *The Land Unknown*, whose strap-line, 'Behind a barrier of Antarctic ice, a paradise of hidden terrors', suggested that all manner of monsters—including the obligatory dinosaurs—still roamed the Antarctic interior. Even more wonderfully, the Antarctic was pressed into service by those demented if ingenious proponents of a flat Earth, such as the notorious 'Parallax'—alias Samuel Birley Rowbotham—in the late nineteenth century and his successor Glenn Wilbur Voliva. For them, Antarctica was not a vast icy island, but a perimeter ring of ice cliffs *to keep the ocean waters from cascading off the Earth.* When Richard Byrd was preparing for the first-ever flight across Antarctica in 1928, Voliva fired off telegrams, warning that he would disappear into outer space. When Byrd achieved his feat, flying over the South Pole, Voliva would have none of it. Byrd, he said, had just circumnavigated the world's edge.[2]

Only since the mid-twentieth century have we begun to understand this continent's nature, its long geological history, and its importance in regulating climate on Earth. This understanding began with the first permanent settlements in the Antarctic: wooden huts, inhabited by bearded and unkempt men. One such 'hut' was on the northern tip of the Antarctic Peninsula at Hope Bay, a locality originally discovered by Swedish explorer Otto Nordenskiöld at the beginning of the twentieth century.

Hope Bay was first peopled in 1945 as part of Britain's mission to strengthen its territorial claims on the Antarctic Peninsula. The conditions that these early 'settlers' endured were primitive. Living in a building that was not far up the evolutionary ladder from a garden shed, though rather grandly called Trinity House, and cut off entirely from the outside world during the long, dark winter months, the men subsisted on a dubious selection of tinned and concentrated food (incidentally, the British Antarctic Survey still send their scientists into the field with this dehydrated fare). Nevertheless, William Turner, the base commander in 1954, was able to note that 'the French tinned varieties of sprouts and asparagus were very good and much superior to English tinning'.

Hope Bay, or Station D as it was originally known in 'James Bond English', was the starting point of some of the epic Antarctic journeys in the late 1940s. Ray Adie in 1948 sledged from Hope Bay to Stonington base ('Station E'), a distance of several hundred kilometres along the Antarctic Peninsula. Sometime after, Adie and Vivian (later Sir Vivian) Fuchs sledged from Stonington to Ecklund Island—with absolutely no back-up. Perhaps among the last truly great adventures of discovery before the space age, their trips were the equivalent of sledging across the entire UK from John o' Groats to Land's End—a few men and a dog team, carrying all their food and fuel, without any air photos, or maps, across a true *terra incognita*, and often on terribly dangerous terrain.

Stretching from Hope Bay on the furthest northern tip of the Antarctic Peninsula to McMurdo Sound in the East Antarctic, the Antarctic ice sheet is simply enormous. In some places it is over 4 kilometres thick and it covers a greater area than the United States or China. All the ice has formed from snow, and eventually it slides into the sea under its own colossal weight. In some places the ice is over one million years old. Within it is preserved a marvellous record of climate change. Flying over this vast wilderness of white and staring at its ancient contours, there comes a great sense of permanence. Surely, you feel in your bones, something this massive must always have existed. From a human time frame of just a few decades, it is difficult to conceive of it ever changing. Human perceptions, in these matters, can be quite misleading.

The Antarctic ice sheet is complex. It is not just a single massive lump of ice, but is divided into two by the Transantarctic Mountains that span the continent. In scale these mountains are the equal of the Andes, but they are almost entirely buried in the ice. To the east of these mountains lies the massive East Antarctic Ice Sheet, the biggest frozen freshwater reservoir on Earth. It accounts for about 90 per cent of the volume of Antarctic ice. It rests on high ground and air temperatures above it may fall below −70 °C in winter (the coldest ever recorded temperature is −89 °C). To the west of the Transantarctic Mountains lies the West Antarctic Ice Sheet, which is younger than its eastern cousin, and less stable. Much of the West Antarctic ice is grounded well below sea level so that without the ice there would be an ocean here. It connects, though, an archipelago of islands that extends northwards to the Antarctic Peninsula, and also rises on to land over 2,000 metres high, and if it were to disintegrate, it would cause about 3 metres of global sea level rise. The Antarctic Peninsula Ice Sheet is the last of the three great masses of ice. It runs the length of the Peninsula, some 1,900 kilometres. It covers peaks rising to 3,000 metres, from which ice slides into the sea on both sides via a series of large glaciers.

The history of the East Antarctic Ice Sheet stretches back over thirty million years. Nothing of that original ice remains, and there are no confirmed strata in Antarctica that signal its presence. Any such deposits have been obliterated long ago by the bulldozing action of ice, or lie hidden beneath the sea. To see the birth of the Antarctic ice sheet one must travel far away and into deep seas. Beneath those ocean floors, there are fossils of organisms that, even from the remotest depths of the oceans and at a distance of thousands of kilometres, sensed the growing ice—such is the interconnectivity of everything on the surface of the Earth.

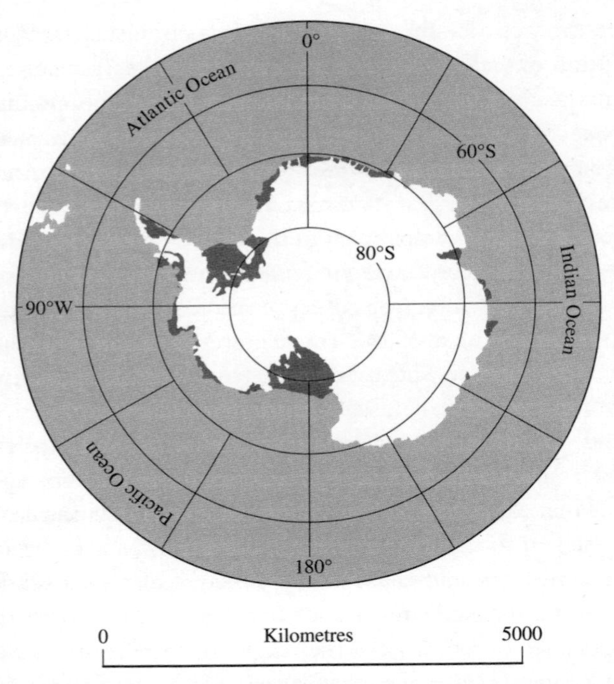

FIG. 16. Antarctica. Areas shown in black are major ice shelves, including those of the Weddell and Ross seas.

Remote sensing the past fifty million years

In the middle of the nineteenth century little was known about the oceans, except that away from the coastline they were clearly very deep. In 1869 French novelist Jules Verne published his spectacular account of the undersea world of Captain Nemo and his submarine *Nautilus*. Readers, then, really could imagine the oceans as frequented by Verne's sea monsters. And yet, just a few years later, in 1872, the Royal Society in London embarked on the *Challenger* Expedition, the first global oceanographic survey. This survey and those that followed would dispel the notion of great sea monsters, and begin the documentation of the deep oceans proper, a documentation that has continued until today. *Challenger* collected thousands of samples on her voyage, among them many foraminifera that are now 'pickled' in Canada balsam, in glass slides secreted in the wooden collection trays of the Natural History Museum in London.

Foraminifera are amoeba-like organisms whose fossil record extends back over 500 million years to the Cambrian Period. Perhaps their success and longevity lie in their simplicity: each foraminifer comprises just a single cell. But these single-celled organisms are also the giants of the 'unicellular world', up to a thousand times larger than a bacterium, more than the difference in size between a man and the Empire State Building of New York. Foraminifera can be thought of as an amoeba with a shell, and that shell takes on a multitude of different shapes and sizes. Some foraminifera build their shells from the sediment around them, gluing this sediment together with 'foraminifer spit' to make a home. Other foraminifera make a shell from calcite, extracting the ingredients to do this from the sea water around them. Foraminifera live on the sea floor, from shallow reefs to the deep oceans, but they also float freely as plankton, building shells that look like a series of interconnected balloons. The most famous of the sea floor-dwelling foraminifera are the nummulites that lived about fifty-five million years ago. Their name is derived from the word for a coin, and they made shells that were absolutely gigantic for a single-celled organism, reaching a centimetre in diameter. Countless skeletons of nummulites make up the limestone used to construct the pyramids at Giza.

The calcite shells of the foraminifera are the key to unlocking the evolution of climate through the past fifty million years, and to reconstructing the growth of the Antarctic ice sheet. How, though, can a foraminifer living on a deep sea floor in the tropics sense ice forming thousands of kilometres away?

The calcite shells of foraminifera preserve a chemical record of the sea water around them. If that record is then locked into the

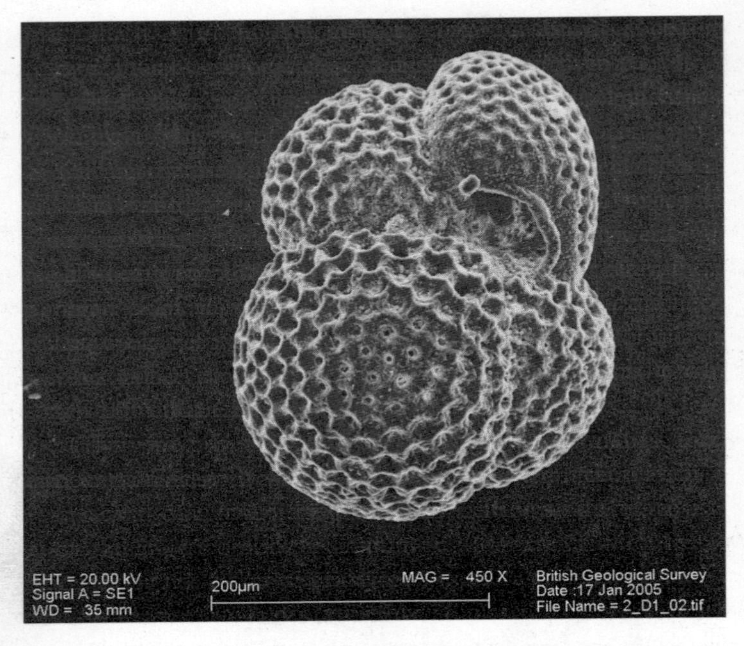

FIG. 17. The planktonic foraminifer *Globigerinoides sacculifer* occupies the surface waters of the tropical oceans. The scale bar is one-fifth of a millimetre.

foraminifer shells—and stays locked in—then a precise record of environmental change can be formed. It is this signal that allows one to sense the growth of the Antarctic ice sheet from the depths of a distant sea floor. But the signal is subtle, and deciphering it needed ingenuity and hard work—and a touch of serendipity too.

Towards the Rosetta Stone of climate

The basic idea came with a throwaway remark after a lecture.[3] The lecture was given by the renowned chemist Harold Urey (1893–1981), a guest in Zurich, and he had been talking about the work for which he had earlier won a Nobel Prize—the discovery that not every atom of an element is alike and behaves in the same way. He had discovered that many elements exist as different isotopes, which, possessing different numbers of neutrons, have slightly different atomic weights—and molecules containing them will have slightly different physical properties. Thus, normal water, say, includes both molecules containing a 'heavy' oxygen atom (^{18}O), and molecules with the 'light' oxygen atom (^{16}O). As water evaporates, the water vapour derived from it will be enriched in molecules with the 'light' isotope, as these are easier to evaporate, and the remaining water will be enriched in the 'heavy' one—an example of mass-dependent fractionation.

Sitting in the audience was a crystallographer, Paul Niggli. Could it be the case, he asked, if crystals precipitate from waters of different isotopic compositions, that the oxygen in those crystals would also reflect those isotopic compositions? And to turn that deduction on its head (for Niggli had the instincts of a geologist) could measuring the isotope ratios of the calcium carbonate crystals of an ancient limestone allow one to say whether they had formed in fresh or salt water?

It was a prescient remark. Urey thought on it, and made some calculations—and discovered that the proportions of the oxygen isotopes that went into a limestone reflected the temperature of the water in which the limestone formed. As the water cooled, then slightly more of the heavy isotope went into the limestone, and vice versa. That was the light-bulb moment. He said later that he had suddenly found a geological thermometer in his hands.

It was a thermometer that needed a lot of hard work to develop, though, not least because the atom-counting machines (mass spectrometers) used to measure the different proportions of isotopes were not yet sensitive enough to detect the very small differences (of less than one part in a thousand) caused by temperature changes in sea water. Nevertheless, as the scientific prize of a functional geothermometer was so great, the mass spectrometers were redesigned ved, and the idea eventually painstakingly tested (the chemical problem he ever faced, according to Urey) and n calcium carbonate mollusc shells grown in sea water at a known temperatures.

Enter, now, a scientist, polymath, and all-round colourful personality—Cesare Emiliani (1922–95), a native of Bologna—to work in Urey's Chicago laboratory. Emiliani liked to shake up the formalities of modern science a little. In one essay that he wrote with six fictitious co-authors (to make up, he said, the mystical number seven), he argued that scientists should no longer write papers for technical journals, but rather should have accounts of their discoveries played by wandering minstrels at night-long banquets. For all those who are a little weary of academic formality, it's a beguiling proposal.

Emiliani analysed the oxygen isotope ratios of the calcium carbonate shells of fossil planktonic foraminifera entombed within ocean floor sediments. Here, he recognized a repeating isotopic pattern of heavier and lighter values that he interpreted as representing alternating colder and warmer temperatures in the seas. Following Urey, he interpreted his isotope data as essentially reflecting the temperature of the water. But, there was another possible interpretation: as the ice sheets grew, they would have formed from snow that had originated as water molecules evaporated from the ocean—and hence would have been richer in the lighter oxygen isotope. So, in growing, the ice sheets would preferentially abstract ^{16}O from the oceans—leaving them richer in ^{18}O; and, as they melted, the ^{16}O would come flooding back into the sea. Perhaps Emiliani's data reflected global ice growth, rather than simply temperature? And, if so, how could one tell?

This question was tackled by another quite distinctively talented scientist (the palaeoclimate community seems to specialize in these): Nick (or, as he became, Sir Nicholas) Shackleton (1937–2006) of Cambridge University, a man who had a remarkably relaxed dress sense, a kindly nature, a passion for playing and collecting clarinets (he amassed perhaps the world's finest collection of these) and a genius for extracting climate data from fossils.[4] Early in his career, he had worked to make mass spectrometers so sensitive that they could extract an oxygen isotope value from just one or two specimens. He turned his attention to the problem of what Emiliani's oxygen isotope signal represented—and solved it with some panache. In the deep sea of today and the geologically recent past, he reasoned, the temperature is always near freezing and, even, during a glacial phase, cannot fall any further. Therefore, any calcium carbonate-shelled animal that lives at depth (such as a sea floor-living foraminifer) will not show any significant temperature change between glacial and interglacial phases—but will show the ice volume signal. So, he compared—from a single deep-sea borehole—the oxygen isotope ratios from both sea surface-dwelling foraminifera that can be affected by glacial–interglacial temperature changes, and the deep-living ones that live forever in cold waters. The patterns shown by both were similar—indicating that the ice volume effect was dominant. Characteristically, Shackleton warmly praised Emiliani's groundbreaking work in his published account,[5] even as he showed his temperature-centred interpretation to be wrong.

The changing chemical compositions of the fossil foraminifera record the changing climate of the Cenozoic, and pin-point the geological instant when a tipping point occurred, and the Antarctic ice sheet first formed. Now that humans had worked out a way to decode the chemical patterns, their story could be told.

Anatomy of a tipping point

The oxygen isotope composition has now been analysed from many thousands of foraminifera from hundreds of deep-sea boreholes that have penetrated the modern ocean floor to reach deep-sea sedimentary deposits dating back to the Cretaceous. In 2001 the palaeo-

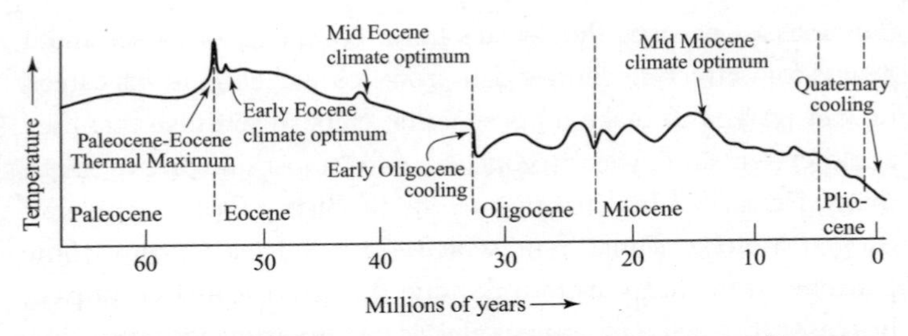

FIG. 18. Major events in the climate history of the past seventy million years. The hyperthermals of the earlier Cenozoic were followed by long-term global cooling. Antarctic glaciation commenced at the Eocene–Oligocene boundary. Extensive Arctic ice sheets were already forming during the Pliocene.

climatologist Jim Zachos of the University of California in Santa Cruz and his colleagues put together many hundreds of man-years of work on these boreholes in one single image. It shows the progression of oxygen isotope values in the ocean from seventy million years ago, within the Mesozoic greenhouse, to the present day.[6]

The image is striking (see Fig. 18). It shows the warmth of the late Cretaceous Period, and that of the early part of the Cenozoic Era. The end of the Cretaceous represented a biological catastrophe, triggered by some combination of meteorite impact and massive volcanic outbursts. However, the long-term impact on global climate per se was slight, and the world after the calamity remained warm. The Zachos curve shows some finer detail, too. There is, as the thinnest of spikes, the brief hyperthermal event at the Paleocene–Eocene boundary—with similar, lesser events later in the Eocene, that overall grew gradually warmer, then cooler. There is an almost-microscopic jitteriness to the line overall, that, if magnified, shows the kind of astronomically related variations that Grove Karl Gilbert had first noticed (Ch. 4) in the rocks of Colorado. Then, exactly at the level where the nineteenth-century geologists had placed the boundary between the Eocene and Oligocene epochs, at what is now calibrated to 33.6 million years ago, the line on Jim Zachos's chart drops like a stone.

It is the biggest single shift in oxygen isotope values in the entire Cenozoic Era, and marks the rapid withdrawal of enormous amounts of water from the oceans, and, for the first time since the ice sheets of the Permian, ice now coalesced in the southern polar landscape to form a major continental ice sheet. How quickly did the Antarctic ice grow? Here, we must turn away from the Zachos composite diagram to the yet more detailed story that an individual deep-sea borehole can tell, when examined closely. This was a little tricky to find, because large areas of the ocean floor at that time were scoured by deep-sea currents, made more powerful as a result of the changes affecting the oceans—and so in effect scrambling or obliterating part of the strata and the evidence contained within them. But, in the central tropical Pacific and just offshore from Africa, there were some calmer areas, and boreholes drilled through these showed the progression of events.[7]

Viewed thus, the rapid change in conditions appears to have been a two-step event, each step being accomplished in a single 40,000 year astronomical 'tilt' cycle, these being separated by a stable climate plateau of about 200,000 years. In all, the sea level seems to have dropped by about 70 metres—suggesting that the Antarctic ice sheet then grew to be much the same size as it is today—while the deep ocean waters cooled by several degrees.

As the oxygen isotopes in the ocean changed, there came the first physical evidence of expansion of Antarctic ice. Sedimentary successions around the Antarctic show, in the earliest Oligocene strata, rock fragments released from floating icebergs. And, such earliest Oligocene ice-rafted debris is recorded as far afield as the Kerguelen Plateau, 1,000 kilometres to the north of the Antarctic continent in the southern Indian Ocean.

The effects rippled across the world. There are strata that include the Eocene–Oligocene boundary in what is now the 'Badlands' of North America, in South Dakota and Nebraska. These fossilized landscapes contain vertebrate remains. Isotopes extracted from fossil tooth enamel suggest that temperatures dropped by some 8 °C—

not instantly, but over about 400,000 years. The mammals coped well, but the reptiles and amphibians, being cold-blooded, suffered.[8] Across the world, on the Tibetan plateau, cooling also took place, and the climate grew more arid, as lakes dried up.[9]

Hence, the growth of the Antarctic ice sheet was linked with changes worldwide, although not—yet—the growth of major ice sheets in the high Arctic. What, though, triggered the Eocene–Oligocene boundary event? While this mystery is not yet fully solved, there are significant clues.

Clues to a global glaciation

In many of the borehole cores pulled up from beneath the ocean floors, the beginning of Antarctic glaciation can be seen as a simple colour change, as dark silica-rich oozes are succeeded by paler oozes dominated by the calcareous skeletons of planktonic foraminifera. This change is not biological, marking a change in the plankton communities high above. Rather, it marks a change in ocean chemistry, showing that the deep ocean waters had become less corrosive to calcium carbonate: the 'carbonate snowline', above which the sinking shells of dead foraminifera are not dissolved, had then suddenly descended, around the world, by about a kilometre.

This is the reverse of what happened at the hyperthermal at the Paleocene–Eocene boundary (Ch. 4). At that event, the temporary rise of the 'carbonate snowline' was likely because of an ocean briefly out of balance, its chemistry overwhelmed by the acidifying effects of the geologically sudden increase in carbon dioxide in the atmosphere. Could this be the opposite effect, caused by a sudden lowering of carbon dioxide in the atmosphere? Well, perhaps in part—but in this case the 'carbonate snowline' stayed low through millions of years of the Oligocene, so an additional mechanism is needed. This is provided by the drop in sea level—of up to 70 metres—that took place as Antarctic ice grew.[10] As the continental shelves of the earliest Oligocene became exposed by decreasing sea levels, limestone could no longer accumulate so widely in shelf seas. In tandem, the rivers, growing longer as land areas increased, eroded newly exposed limestone on the continental shelf—and carried this carbonate into the deep oceans. As the oceans became more saturated in carbonate ions through the Eocene–Oligocene boundary, the depth at which limestones could accumulate in the oceans increased rapidly by as much as a kilometre. With less shallow marine area to act as a reservoir for carbonate, more carbonate had to be stored in the deep ocean—and so the 'snowline' dropped.

But what, then, acted to trigger the growth of ice, the drop in sea level, and the change in ocean chemistry? Perhaps the redistribution of heat around the globe was an important factor, with ocean currents being forced into different patterns as the shifting continents changed their positions.[11] During the Cenozoic, there were major changes in the configuration of the connections between ocean basins, as high-latitude ocean gateways opened (with the separation of South America from Antarctica), and low-latitude gateways closed (as Africa closed towards Europe and Asia, and later as the Central American Isthmus formed). Those processes continue to act today. Thus the annual influx of surfing and holidaying Australians to the beaches of Bali is, in its way, a kind of premonition: as the tectonic plate that contains Australia and New Guinea drifts ever northwards, it will eventually collide with south-east Asia. These changes, beginning in the Eocene, probably contributed to the development of the Antarctic circumpolar current. They forced a dividing line in the sea, beyond which no warm surface water from lower latitudes could flow south.

However, computer models of the beginning of glaciation suggest that changing the Earth's heat distribution patterns would have been insufficient in itself to trigger the growth of ice on a massive scale.

These models point the finger at carbon dioxide, and suggest that there must have been a reduction in the concentrations of this greenhouse gas in the atmosphere around the end of the Eocene and the beginning of the Oligocene.[12] At some point, these models suggest, a threshold level was reached that could trigger sufficient cooling to push Antarctica into a glaciated state. Climate models of the southern polar region at this time show that the ice may have formed rapidly once carbon dioxide levels dropped below a threshold value of 2.5 times its present level. Is there any evidence, though, of what carbon dioxide levels were doing at that time?

The carbon dioxide drawdown

There is no fossil air from the mid-Cenozoic, more's the pity—and estimating ancient carbon dioxide levels of this antiquity is very difficult. However, the same fossilized skeletons of foraminifera that allow us to gauge past sea temperatures also provide some idea of the acidity of these ancient seas—and from that information comes a sense of changes in the carbon dioxide in the atmosphere above. This kind of analysis has been done using the stable isotopes of carbon and also of the element boron, which is used, for example, to temper glass and ceramics. Boron has two naturally occurring stable isotopes, ^{11}B and ^{10}B. The way in which these isotopes move from sea water into marine carbonates is directly determined by the acidity of the water in which they are dissolved. As acidity in turn is a function of carbon dioxide in the atmosphere and oceans, the ratio of ^{11}B to ^{10}B in foraminifer shells has been used as an indicator of past ocean acidity and of atmospheric carbon dioxide.[13]

Such measurements show that the Cenozoic Earth seems to have undergone a long-term process of significant decrease in atmospheric carbon dioxide levels.[14] During the peak warmth conditions of the mid-Eocene Epoch, about fifty million years ago, levels were probably around 1,000 ppmv. By the Oligocene Epoch, some twenty million years later, the chemical data suggest that levels had fallen to much lower carbon dioxide concentrations, of some 500 ppmv. These values are consistent with the modellers' suspicion that carbon dioxide was the driving force behind the refrigeration of climate. But what drove down carbon dioxide levels?

The growth and destruction of mountains

Strata contain many patterns that act as indicators of wider events in the world around them. But, while the patterns might be clear and reproducible, the events that have produced them often remain shadowy. One such pattern was, over thirty years ago, implicated in the cooling of the world—yet, even now, its significance and role remain uncertain.

The pattern is that of varying proportions of two isotopes of the element strontium, which is a common dissolved constituent of sea water. As we have already seen, one of these, the 'heavy isotope' ^{87}Sr, is more commonly sourced from continental rocks; the other, the 'light' isotope ^{86}Sr, is more typically derived from the ocean crust (see Ch. 3). In strata laid down over the last sixty million years, there has been a clear, quite undoubted trend towards increasing amounts of 'heavy' strontium.[15]

There seemed to be a ready explanation for this. India, which had been drifting northwards, collided with Asia in the early part of the Cenozoic Era, the collision producing the mountain belts of the Himalayas and the Karakoram. Therefore it was suggested, quite reasonably, that the growth of such young mountain belts, most particularly the Himalayas, exposed huge amounts of fresh rock surface to the atmosphere, which would have undergone rapid weathering, drawing carbon dioxide out of the atmosphere in the process. If enough 'extra' mountains are weathered in this way for long enough, then carbon dioxide levels in the atmosphere will drop to the point that glaciation can begin.[16]

It is a very neat and simple idea, and one, that as we have seen, has been applied to other and earlier rock successions—and it may have been a factor in this case. But it has been frustratingly difficult to pin down—despite (or perhaps *because*) the evidence in the Cenozoic is more plentiful than in the more ancient examples.

The erosion of mountain belts is a physically and chemically complex process. Much of the detritus is simply physically disaggregated, and then washed downriver to the sea as chemically unweathered grains—which cannot then contribute to natural carbon dioxide sequestration. And, when mountain belts erode, some of the rocks are carbon-rich, and part of this, on weathering, re-enters the atmosphere (while another part is simply washed as particles back into the sea and buried again).

Some rocks, though, are more effective at capturing carbon dioxide than others. Rocks of the ocean crust—especially the heavy 'ultramafic' rocks in its lower part—are rich in calcium and magnesium silicates. These therefore draw a lot of carbon dioxide out of the air on being weathered, the original minerals converting to calcium and magnesium carbonates. Such rocks do not normally get to the surface where they can be attacked by the weather, for they lie at the bottom of the ocean and, when their surface existence comes to an end, they slide back down into the Earth's mantle at a subduction zone. By a quirk of plate tectonics, though, a large slab of Pacific ocean crust, of some tens of thousands of square kilometres, escaped this fate near the end of the Eocene Epoch. It was flipped up, then pushed across the tropical island that we now call New Caledonia. Chemical weathering of this single gigantic crustal slab might have drawn enough carbon dioxide out of the air to trigger the Cenozoic ice age.[17]

As so often with science, there are too many suspects, and still too little hard evidence, to settle on the main culprit. We are left with that astonishing two-step drop that changed the Earth's climate regime (the pattern in itself suggests that the process of cooling was paced in a quick-slow-quick fashion by astronomy, by those Milankovitch patterns of the Earth's orbit and spin). There remains the deep suspicion—with some evidence to support it—that carbon dioxide levels played a major role, as did also the Earth's heat transport mechanisms. From that time, the Earth was once more in icehouse mode.

Following the step change, some thirty-three million years ago, that equates to the Eocene–Oligocene geological boundary, climate proceeded in a succession of lurches, separated by more stable intervals, as the Zachos diagram clearly shows. Another change is visible at the Oligocene–Miocene boundary twenty-four million years ago and yet another, fourteen million years ago, at the boundary between the middle and late Miocene.[18] The last of the big step changes occurs about two and a half million years ago and defines the boundary between the Pliocene and Pleistocene epochs. Each of these changes is a tipping point in the Earth System, at which some combination of factors—greenhouse gases, ocean gateways, rising mountains—interacted to take the Earth, each time, into a different climate state, usually a colder one. We need to take a closer look at some of these step changes, to get a sense of how the Antarctic and Arctic wildernesses emerged.

The isolation of Antarctica

For many travelling with the British Antarctic Survey to the Antarctic Peninsula in the 1990s and 2000s, sailing across the Drake Passage on the Royal Navy's HMS *Endurance* was the usual port of entry. This great ship, bearing the most famous name in Antarctic exploration, nearly sank on its way back across the sea in 2008, with the crew and scientists forced to bail out water using a train of buckets as it limped into Ushuaia in southern Argentina. With no decision yet about refitting, *Endurance* now lies in dock in Portsmouth, the victim of UK government cutbacks and a ghost of her former self.

Those lucky enough to sail with *Endurance* encountered the journey across the Drake Passage that sailors have always regarded with awe. Once out of Mare Harbour, the Royal Navy's base on the Falkland Islands, the seas first appear becalmed. But from then on, even the greatest of Antarctic scientists have often spent much of the journey across the Drake Passage pinioned in their tiny bunk, dodging the flotsam of the cabin and desperately trying to retain their last meal as the ship pitches and lunges. Endurance is a shallow-drafted ship, designed to rise up over the ice and break it. But as the crew would happily tell you, when she hits a storm her flat bottom makes her 'roll like a pig'. Not until halfway across the Drake Passage when the great bergs of ice first appear on the horizon, can one wonder at the land beyond, the most remote wilderness on Earth. The seas of the Drake Passage, then, are fickle, and often ferocious. At this latitude, between South America and the Antarctic Peninsula, there is no significant land to the west or east around the entire planet. If latitude forties winds are roaring, then those of latitude sixties live up to their name as 'screaming'.

The seas in the south were not always so fearsome. Before forty million years ago, South America and the Antarctic Peninsula were joined. But these crustal masses have since parted, separated by plate tectonic processes deep under the Southern Ocean. Such a process may have begun some thirty-five million years ago, but the passage between these lands had become a deep ocean by the early Miocene, twenty-four million years ago.

In tandem with the opening of the Drake Passage, plate tectonics also caused Australia to drift away from the Antarctic, perhaps breaking that land away as long ago as the Cretaceous. Certainly, by twenty-four million years ago, there was a considerable stretch of deep ocean between these lands. It is called the Tasman Gateway.

The break-up of the remaining land mass of Gondwana into the southern continents is suspected of driving the dramatic effect on Antarctica's climate. First it allowed the establishment of the Antarctic Circumpolar current that runs unimpeded around the globe at 60° S. The rocks beneath the Tasman Gateway still record the formation of this current. Drilling the ocean bed here reveals tell-tale signs of an ocean current accelerating in the earliest Miocene, twenty-four million years ago.[19] At this time there is a sudden increase in size of the sediment grains preserved in the strata, signalling a strengthening current. The Antarctic Circumpolar current effectively cut off the supply of warm surface water from lower latitudes, triggering the 'screaming sixties'. This amplified the cooling effect and the Antarctic was plunged deeper into the cold. Weathering and uplift of the Himalayas also seems to have reached a peak at this time. The Earth cooled, and Antarctic ice spread further.

The last of the Antarctic tundra

In the Cenozoic the Antarctic witnessed an interlude of relative warmth during the 'Mid-Miocene climate optimum'. This climate optimum reached its peak some sixteen to fourteen million years ago, long after the Antarctic Circumpolar current had formed, and long after the initial drop in carbon dioxide that probably precipitated the formation of the polar ice-sheet at the Eocene–Oligocene boundary. There is evidence in the shells of sea-bottom foraminifera that the global climate warmed during the Mid-Miocene climate optimum and some have linked this warm interval with the volcanic rocks of the Colombia River flood basalts. The outpouring of these volcanic lavas, one of the largest to have occurred during the Phanerozoic, covered great areas of Washington State, Oregon, and Idaho and would have released much carbon dioxide to the atmosphere. Perhaps for a time, this may have produced a temporary respite from the removal of carbon caused by the rising Himalayas. Far away in the remote areas of Antarctica the local climate responded to this global warming. The fossil evidence of this is recorded within the Dry Valleys.[20]

The Dry Valleys region of the Antarctic is one of those few places on the continent where the ice cover is thin or absent, and where geologists can get to grips with the rocks and the more ancient history of that land. We have already visited this place when we debated life on a Snowball Earth (Ch. 2) and found lakes persisting below thin Antarctic ice in dim light, yet still harbouring life.

There is evidence here in the Dry Valleys of an ancient lake: one that teemed with an assemblage of animals and plants much more diverse than in any modern Antarctic lake. A few years ago, a North American team led by Adam Lewis, Allan Ashworth, and David Marchant found, on the flanks of Mount Boreas, a small outcrop of rock that preserved an ancient lake bed from fourteen million years ago. They could date the lake because nearby volcanic ashes yielded crystals that could be radiometrically dated. In this lake had once lived aquatic mosses, tiny crustaceans called ostracods (more commonly known as seed shrimps), molluscs including snails, and tiny diatoms—photosynthesizing algae that produce minute shells of silica. Around the lake lived beetles, and there were shrubs of the southern beech too, the brown autumnal hue of its fallen leaves still visible. The fossils are preserved in intricate detail: even to the legs and other soft parts of the ostracods, and there is no doubt therefore that the animals and plants lived and died in the lake. But none of these animals or plants live in the Dry Valleys today. And indeed, there are no ice-free lakes that could allow such animals to get a toehold. Summer days in the Dry Valleys are typically no warmer than −12 °C while the winter nights are bitterly cold.

And yet, fourteen million years ago, this place was a thriving tundra landscape, like the steppes of northern Siberia. That is a harsh enough climate in itself, but it is a good deal warmer than the intense polar desert of the Dry Valleys today. The animals that lived in this lake fourteen million years ago provide a precise measure of the change in climate between then and now. The tiny ostracods are miniature cousins of crabs and lobsters. Swimming and darting among the pondweed of any lake or pond at lower latitudes, ostracods are not found anywhere in the Antarctic continent today, but they are found in the lakes of sub-Antarctic regions such as South Georgia, where the mean summer temperature is above 1 °C, and lakes are therefore seasonally free of ice.

Between now and then, the mean annual temperature in the Dry Valleys has dropped dramatically. And the evidence from the ocean basins suggests that there was a rapid deep freeze after fourteen million years ago. As climate cooled through the late Miocene Epoch, vegetation changed not just in Antarctica, but everywhere. And so Africa became more arid and in North Africa the tropical vegetation of the early Miocene yielded to the Sahara Desert of the latest Miocene. As grasslands spread across Africa, these changes had a considerable impact on primate evolution: our last common ancestor with chimpanzees is traced back to the late Miocene.

Opposing Antarctic histories

High in the Transantarctic Mountains, fossils of marine diatoms may be found in glacial sediments that are dated as Pliocene in age, some three million years old. These diatoms originally lived in the seas that filled the Wilkes and Pensacola Basins during an ice-free period of the Pliocene. As they died, they sank and were preserved in the sediments at the sea bed. As climate cooled again, the ice reinvaded the sea, scraping up and carrying the sediments far away to deposit them in the Transantarctic Mountains, where they are found today. The diatoms have formed the basis of an argument that has raged for three decades now, about the stability of the Antarctic ice sheet.[21] The opposing camps comprise the 'stabilists'—who argue for a stable ice sheet since the late Miocene, and the 'dynamists'—who maintain that it has changed substantially. This debate is important because deglaciation of just part of the Antarctic (or Arctic) ice sheets could produce a sea level rise sufficient to flood many of the world's great coastal cities.

Locking up 70 per cent of all fresh water on Earth, the East Antarctic Ice Sheet would cause 60 metres of global sea level rise if it were to melt, and it helps regulate the Earth's climate by reflecting heat back out to space. Those who favour its essential stability have used the Dry Valleys region to suggest no change for the past fourteen million years. Indeed, the landscape evidence from the Dry Valleys suggests no fundamental change in climate since the cooling at the end of the Middle Miocene that obliterated the lake fauna and flora of Mount Boreas. The 'dynamists', by contrast, argue that the Dry Valleys is a special place with its own microclimate, and thus not representative of the East Antarctic Ice Sheet as a whole. In support of this hypothesis, some of the sea level changes estimated for the Pliocene that exceed 10 metres would need a significant loss of this ice sheet.

At Oliver Bluffs in the East Antarctic, just 500 kilometres from the South Pole, another fossil locality preserves a wonderful plant fauna that contains the leaves of the Southern Beech tree, preserved with their brown autumnal hue, just as they are at Mount Boreas. Diatoms from this locality also suggest a Pliocene age, and if this were so, conditions here must have been much warmer and the East Antarctic Ice Sheet somewhat smaller, and thus more dynamic, than today. But the dating of these deposits remains contentious, for there are no volcanic deposits here to give an absolute age. More strikingly, in the warmer Antarctic Peninsula region there is no evidence for tundra organisms returning during the Pliocene, even during the warmest intervals.

To add fuel to this debate the stabilists have argued that these diatoms high in the Transantarctic Mountains are wind-blown and sourced from younger deposits elsewhere in the Antarctic. And they have claimed they can prove this because the diatoms are found only on the surface of the rocks in the mountains, and are not found as you dig deep below the surface.

If the East Antarctic Ice Sheet has—arguably—long been stable, then a different story emerges from West Antarctica. Towards the end of the Cenozoic Era, in the Pliocene Epoch between some five and three million years ago, large ice masses grew and collapsed repeatedly. This is significant. In the twenty-first century atmospheric carbon dioxide levels will almost certainly return, for the first time in three million years, back to the levels of the Pliocene. It is to this epoch that we now turn, to seek some kind of picture of the near future.

6 The Last of the Warmth

The Pliocene Pan-American

The Pan-American Highway rises in the far north of the Americas at Prudhoe Bay, Alaska and, except for a small gap in Panama, runs the entire length of the two American continents to terminate at Ushuaia in southernmost Argentina. Along its way it travels nearly 50,000 kilometres, from the polar landscape of the far north, through the boreal forests of Canada, the temperate plains and hot deserts of the USA and Mexico, and on further into the tropical zones of Central and South America, until it reaches the sub-polar landscape of Tierra del Fuego.

The American landscape was not always like this. To travel along the Pan-American Highway some three million years ago, in the Pliocene Epoch, would have revealed a different world. It was a little warmer than our own. Far away, the Greenland ice sheet covered only a small part of that land mass. At the other end of the world, there was less ice covering the West Antarctic than we are familiar with today.

Going south, from Prudhoe Bay along the Pan-American Highway of the Pliocene, there was none of the scrub tundra now seen by the ice road truckers. Forests then extended far to the north, covering vast areas of northern Canada and Alaska, and draping the coastal margins of Greenland. They stretched, too, into Siberia, a mass of forest extending thousands of kilometres from Norway to Kamchatka. There was almost no tundra in the north, except for a few patches in Greenland and on the far northern extremities of Siberia. Instead the polar sun rose across that well-nigh endless green Pliocene forest.

Such a prehistoric journey south along the Pan-American Highway would take one across the grasslands of temperate America. These are truly ancient, having been long established even then. Patterns of seasonal temperature and rainfall, though, allowed forests to grow where none are present today. There were no humans to cut down the trees or hunt the animals that lived in the forests. There were no Great Lakes either, for no northern ice had grown yet, to scour out their floors and fill them with melt water.

Moving further south, towards Central America, the road would stop abruptly. One could go no further at that time. The land that forms the modern countries of Costa Rica, Honduras, and Panama was not yet constructed. It was a chain of islands rising in the sea. The tectonic plate of the Pacific, a slab of rock some 200 kilometres thick, was being thrust beneath that of its Atlantic neighbour, and the pressures generated were forcing the ocean floor upwards, while volcanoes erupted amid this newly risen crust. The animal populations of North and South America were still isolated.

To travel further south, one would have to take a ferry between the islands until land was picked up again, to get into South America proper. There, the tropical forests of South America were already ancient, but they were not yet populated by the jaguar, for that animal only arrived here from the north after the Americas had joined.

The continuing route would skirt the rising Andean highlands, and then the high peaks of the Patagonian Mountains, already topped with small Alpine-type glaciers. Finally, one would reach the southern tip of the Americas; to modern inhabitants of Ushuaia, the climate would feel distinctly warm.

Why does the Pliocene matter?

Such a prehistoric trans-Americas journey shows the Pliocene world as different from our own. And that is why the Pliocene really matters, because something like it may be where we are all heading. The journey to the past may—in some respects—be eerily similar to a journey into the future.

The Pliocene geography, though, was becoming similar to our own. The Himalayas were already fully established. Antarctica was separated from Australia and New Guinea by the widening Tasman Gateway, and from South America by the windswept Drake Passage. A recognizable North and South America had almost joined with each other at the Central American Isthmus. New Guinea was pressing northwards on South-East Asia, and in Europe, the Mediterranean was being squeezed from the south as Africa moved north, slowly encroaching upon Europe. The dominant patterns of modern ocean–atmosphere circulation were establishing themselves, from the Gulf Stream in the north Atlantic to El Niño in the Pacific.

The Pliocene is also the only time in the previous three million years that we have evidence for atmospheric carbon dioxide levels as high as 400 ppmv:[1] nothing in the climate records of the seabed, from fossil plants or from Antarctic ice, suggests levels reached this high again in the intervening time until the present (see Ch. 8). Not until Abraham Darby began to burn coal on an industrial scale in the early eighteenth century did carbon dioxide levels begin to rise again towards levels that are, as we write, within a whisker of those of the Pliocene—and set to overshoot them handsomely.

Comparing the Pliocene and present-day worlds is instructive. Computer models used to reconstruct Pliocene climate show average global surface temperatures warmer than today by about 3 °C, though the regional picture is much more complicated.[2] Pliocene surface temperatures, thus, are within the range estimated for the late twenty-first century, when carbon dioxide in the atmosphere is predicted to double pre-industrial levels. The Pliocene looks more and more like a potential late twenty-first century climate scenario on Earth. That is why some of the best climate scientists have focused their attention on to it.

The Pliocene viewed through a PRISM

About twenty years ago, a small group of geologists of the United States Geological Survey set about reconstructing the climate of the Pliocene world. Why the Pliocene? They had been influenced by that great Belarusian climatologist Mikhail Budyko, who had already written in his 1982 book, *The Earth's Climate: Past and Future*, that the late twenty-first century climate might produce a world not unlike that of the Pliocene.

The USGS scientists scanned the geological records preserved in the rocks of the ocean basins, and happened upon an interval of time within the Pliocene that was about 300,000 years long, and set about three million years in the past. They did not choose this time arbitrarily. They recognized that within the long climate cooling trend of the Cenozoic, and of the Pliocene itself, the rocks some three million years ago recorded the last gasp of sustained warmth before the intense bipolar glaciation of the Quaternary set in.

The USGS team set out their ambitious task: to reconstruct the whole world climate three million years ago, as completely as possible. Not just for the land, but for the shallow seas, the deep seas, the polar ice sheets, and the atmosphere. For this ambitious task they invented the name 'Pliocene Research, Interpretation and Synoptic Mapping'. It didn't quite possess the elegance of a Shakespearian sonnet, though it neatly synthesized the main components of the work. For brevity, they settled on PRISM. A new geological acronym was born and, quirkily, the Pliocene world could now be viewed through a PRISM.[3]

The idea of reconstructing the Pliocene world of three million years ago might have seemed, then, an insurmountable task. Where was the geological evidence to do this? Significant amounts of information about the Pliocene world had been accumulating for over 150 years, to be sure. Indeed, the name 'Pliocene' had been coined by that nineteenth-century scientific giant, Charles Lyell, geological mentor to, and a close friend of, Charles Darwin. On the English coast of the North Sea, geologists working in the Crag (shelly sand) deposits of East Anglia had already recorded the presence of 'Mediterranean'-looking fossil molluscs, which suggested to them somewhat warmer conditions in the Pliocene North Sea.[4]

What followed then from the PRISM group was a mammoth task of correlation, to make sure that the Pliocene deposits from an ancient forest in northern Canada were of the same age, as nearly as possible, as Pliocene deep sea foraminiferal oozes from the bottom of the Atlantic Ocean, as shelly sands from around the North Sea, and so on. Only when this correlation of strata and the events preserved in them was worked out properly around the world, could ancient climate reconstructions be made with any degree of certainty. PRISM had to build a map of the Pliocene world.

At first, PRISM concentrated on reconstructing the sea temperatures of the north Atlantic. There was evidence of Pliocene warmth from those fossiliferous Crag sands of eastern England, but also from the fossil mollusc faunas of Iceland and Greenland, suggesting that the north Atlantic was much warmer in the Pliocene. The man charged with the task of exploring Atlantic warmth was a young palaeontologist specializing in foraminifera called Harry Dowsett. Back in 1990, Harry was a recently emerged Ph.D. student from Brown University. He set about his task with notable energy.

Harry established a relationship between modern planktonic foraminifera in the surface oceans, and the patterns of fossil planktonic foraminifera preserved in rocks of Pliocene age in the same seas. By using boreholes drilled into the ocean floors, he could compare those foraminifera retrieved from the surface layers of sediment, forming today, with those tens to hundreds of metres physically lower and three million years further back in time, in the same borehole. He was able to use the temperature tolerances of the modern organisms to identify changing patterns of surface temperature between the present and the Pliocene.

The fossil foraminifera acted as a temperature gauge. The fantastically named *Neogloboquadrina pachyderma* was, and remains (it is still alive today) a cool polar form, while *Globigerinoides sacculifer* (also a survivor from the Pliocene) is a seeker after the good life, found basking in the surface waters of the tropics. By 2010, Harry would have been a rich man if he had earned a dollar for every fossil he picked, analysed, and counted: he passed the million-fossil mark that year.

The foraminifer patterns indeed showed warmer conditions for the north-east Atlantic.[5] Harry and his colleagues suggested that enhanced surface ocean circulation from the tropics to the high latitudes may have delivered more heat to the north-east Atlantic, but they were not yet sure of the controlling mechanism. Was it ocean currents or was it elevated levels of carbon dioxide in the atmosphere that gave the warmth?

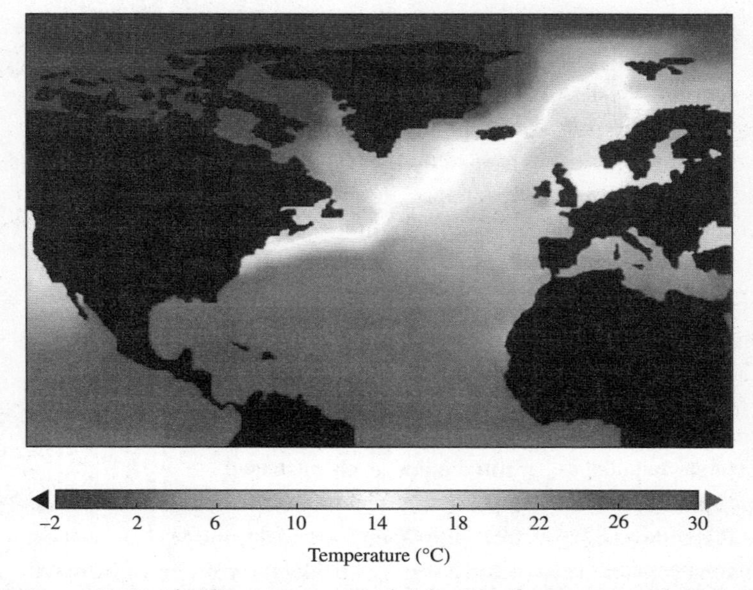

FIG. 19. During the Pliocene, warm North Atlantic surface water extended into higher northern latitudes than at present.

Reconstructing the atmosphere of the Pliocene

The leaves of oak trees, and indeed of all broad-leafed plants, bear stomata, tiny openings on the leaf that allow the transfer of gases and moisture between the plant and the air around it. For many plants there appears to be a direct relationship between the number of stomata on the leaf and the amount of carbon dioxide in the atmosphere. When carbon dioxide is high there are fewer stomata, and vice versa. Some adventurous scientists have tried to use this relationship to infer atmospheric carbon dioxide levels in very ancient plants, going back to the Carboniferous. It is easier, though (albeit still a little contentious) to apply this technique to the Pliocene, because oak trees were growing widely at this time and their fossil leaves are well preserved.

Over the past 200 years, as atmospheric carbon dioxide levels rose, the density of stomata on the leaves of many shrubs and trees has shown a marked reduction. In 1996 a young German scientist, Wolfram Kürschner, wrote an influential paper reconstructing carbon dioxide levels from fossil oak tree leaves for a series of time intervals within the past ten million years.[6] To check this trend, he first calibrated the carbon dioxide response of his chosen oak trees from modern examples. He showed that sunlight and moisture levels could affect the number of stomata, but he was able to distinguish between these effects. There was a clear pattern: the number of stomata fell as carbon dioxide levels increased.

Kürschner then applied this technique to fossil Miocene and Pliocene oak leaves from the Open Cast coal mine at Hambach on the Rhine. He recognized a series of peaks and troughs of stomatal density that he recalculated to carbon dioxide levels varying between about 250 and 380 ppmv. One of these peak levels of carbon dioxide, about three million years ago, equates to our Pliocene warm interval. Since Kürschner's pioneering work, other geological proxies for ancient atmospheric carbon dioxide have been developed, including boron isotopes (see Ch. 5). One can also measure the composition of complex organic molecules called alkenones that are produced by marine photosynthesizing algae; these give a signal both of sea temperatures and of dissolved carbon dioxide in the sea. These proxies have broadly supported Kürschner's conclusions.

The impact of elevated carbon dioxide on climate model reconstructions of the Pliocene is profound. The Earth warms everywhere from the tropics to the poles, but the polar warming is always particularly marked. That cannot be the whole story, because the interplay of many different factors contributed to Pliocene warmth.[7] The forests and ice also control albedo, and ocean gateways—narrow sea passages between land masses—influence ocean circulation and hence the distribution of warmth and atmospheric moisture around the world.

Building models of a Pliocene world

The British are obsessed with weather. Any conversation struck up with a local will involve some commentary on the possibility of rain—even on a beautiful summer's day. Partly, perhaps, because of this obsession, the British have built for themselves one of the great state-of-the-art weather prediction centres in the world. It is called the Hadley Centre, named for the great George Hadley, the eighteenth-century lawyer and (amateur) meteorologist who discovered the mechanism that drove the Trade Winds, recognizing the importance of the Earth's spin in driving east–west winds. Great banks of computers line the walls of the Hadley Centre. They have the kind of computer capacity you would expect on the deck of a science-fiction spaceship. They make fifteen billion calculations per second. They need that power because of the quirky nature of the British—and indeed of global—weather.

While the massive computers of the Hadley Centre can model the weather coming for the next few days, their computing power can also be garnered to look back into deep time (see Ch. 2), into the Pliocene and into still more ancient intervals of time. And so it was this powerful modelling facility that a young man called Alan Haywood intended to utilize as he set off on an improbable mission, to reconstruct the climate of the Pliocene. Alan might well have set himself a five-year mission. He was certainly going to be seeking out new life, even if it was in a fossilized form. And for that he would need to chat to Harry Dowsett—and thus a great scientific partnership was born. The synergy between Harry and Alan has produced one of the most effective reconstructions of deep time climate: Harry, the palaeobiologist, providing the data for the boundary conditions that constrain Alan's climate models.

The computer models have become increasingly complex, and the data that underpins them ever larger. Great forests of the Pliocene have been reconstructed from the fossils preserved in terrestrial Pliocene rocks. Hundreds of sites in the ancient oceans have been logged, from top to bottom, using the fossil foraminifera preserved in Pliocene marine strata. And, the ice sheets have been modelled and those models tested against fossil data from the Arctic and Antarctic. Like a laboriously reconstructed Humpty Dumpty made from many, many parts, the models can now examine the interaction of atmospheric greenhouse gases, ice sheets and albedo, forest cover, altitude, ocean circulation, and a plethora of other data.[8] They give telling insights into the past. They show the polar ice sheets, especially Greenland, as much smaller than today. They show the temperatures at high latitudes, both Arctic and Antarctic, to be much higher than present, while forests covered the northernmost lands of the northern hemisphere. One of the best places to chart the distinction between the Pliocene and the present is in the far south, in Antarctica.

The Antarctic Peninsula

The Antarctic Peninsula is a remote place. Indeed, visiting this place in the early twentieth century was almost the equivalent of trying to land people on Mars today. This region nearly did for Ernest Shackleton's Antarctic expedition in 1914, when the original timber-framed *Endurance* was trapped and then sunk by Weddell Sea ice. Shackleton's men finally found refuge on Elephant Island off the northern tip of the Antarctic Peninsula, and Shackleton and five companions made their epic voyage across the Drake Passage and turbulent Scotia Sea to South Georgia, and took refuge at the whaling station there. The harshness of the climate meant that there were no permanent scientific bases in the Antarctic in the early twentieth century, and no way of maintaining the logistical support that such a station would need. There was, therefore, no means of recording the long-term climate of this region.

During the Second World War the Antarctic was finally brought under the umbrella of permanent human occupation. Fears that the Axis Powers (or indeed the United States) might secure a toe-hold in the Antarctic Peninsula were dispelled when in the austral summer of 1944 a team of British 'colonists' (they were in fact commandos) was sent 'south' to found a base on the Antarctic Peninsula. Supported by the ships SS *Fitzroy* and HMS *William Scoresby*, the colonists made land at Port Lockroy ('Station A') in the late afternoon of 11 February.

Six days later the two support ships sailed away and Britain's first permanent Antarctic base was born. In the first three months the men worked hard to build the 'base hut', impressively christened Bransfield House. This establishment of Britain's first Antarctic base was called Operation Tabarin, and the rocks and fossils collected by those early 'colonists' are still visible in the collections of the British Antarctic Survey (BAS) in Cambridge. The BAS grew out of this military operation and in its early years the weather-beaten scientists that ploughed the Antarctic wastes in the 1940s and 1950s still resembled the original commandos, balaclavaed and with ice-rimmed beards, living in the wood huts of Hope Bay Station and others (Ch. 5), while the outside temperature sank to well below freezing. The establishment of a permanent presence in the Antarctic provided a means of tracking climate change.

One of the places on Earth that has witnessed the most rapid regional warming since the 1950s is the Antarctic Peninsula. That opening scene of the Hollywood blockbuster 'The Day After Tomorrow' begins with an ice shelf collapsing into the sea. It looks far-fetched (and much of the movie that follows certainly is so), but an ice shelf two-thirds the size of Belgium really did collapse into the Weddell Sea over a couple of weeks, during the late austral summer of 2002. This was the 'Larsen B' ice shelf—named for the same Carl Anton Larsen who discovered Seymour Island in Chapter 5. There are many pictures of Antarctic scientists sledging across this ice shelf in the twentieth century, but the area now has returned to the sea.

Since the 1950s the surface air temperature of the Antarctic Peninsula has warmed by about 3 °C. The sea ice is much less extensive now, and if the thermostat of the Peninsula is now being reset, then clearly this is a place where we should try to measure ancient climate change.

For geologists exploring the past climate of the Antarctic Peninsula during the warm Pliocene, one place is special. This is James Ross Island. Lying near the tip of the Antarctic Peninsula at 64° S, this island is a volcanic edifice within an icy sea. Its main volcano, Mount Haddington, is buried in an ice sheet 400 metres thick. Here and there, though, scattered along the rugged island coast, rocks emerge through the ice and snow and reveal a story of ancient climate.

To work in this landscape is quite magical. To stand on the ice-covered shore of James Ross Island, to peer through the bergs and wave goodbye to HMS *Endurance* as she sails away on missions elsewhere, gives an experience of utter remoteness. Once gone, you are alone, with just the sound of the wind and your field assistant for company. The pyramid-shaped tents that BAS still uses are direct descendants of those used by Captain Robert Scott (and the New Zealanders still call them Scott tents). They can withstand winds of 100 kph—though Antarctic veterans tell of the metal tent poles being bent double as the intense katabatic winds of the Antarctic blow.

Inside a BAS tent there is room for two to sleep. A line of boxes for food and cooking utensils down the middle provides the only division of personal space. Socks hung to dry in the tent are warmed by the heat and light of a kerosene Tilley-Lamp. A paraffin stove cooks your porridge in the morning and a dehydrated curry in the evening. There is a bottle of whisky in your pack, to warm a cold soul after a day hunting rocks in the field.

The rocks of James Ross Island are dominantly volcanic. They rise high from the sea as cliffs, above which lies the James Ross Island ice sheet. In the warm summer months the ice melts and waterfalls pour down the cliff face. The volcanic rocks of James Ross Island formed over the last ten million years and were often erupted underneath ice. By looking at how the volcanic eruptions were affected by the mass of ice above, one can reconstruct, even now, millions of years later, the thickness of the ice that once enveloped them.

The scientist who has studied these ancient volcanoes in great detail is John Smellie, a native of the small fishing village of Port Seton, east of Edinburgh. John has long been one of the chief scientific explorers of the British Antarctic Survey. Like a latter-day Ernest Shackleton, he has visited the Antarctic twenty times and, not to be outdone by anyone else, has also notched up eight trips to the Arctic. His fame in polar circles is well established and he will forgive us for revealing that he has a mountain named after him in the Antarctic, called 'Smellie Peak' (a Scotsman, John's surname recalls the famous spy of John Le Carré novels, being pronounced Smiley). Once on a sledging trip out of a (by then) deserted Hope Bay he and a colleague were caught in a severe storm on the Trinity Peninsula. Sleeping, they awoke to find a hole ripped in their pyramid tent. On going outside they could see the wind bombarding the tent with razor-sharp ice crystals, shredding it. John was a young explorer then. Alarmed, he asked his colleague Ian Lovegrove what their chances of survival were, in the wilderness, 700 kilometres from the nearest help. Ian replied with the experience of an old hand 'Don't worry John, we're nowhere near losing our lives yet but we're very close to having a very bad time.'

FIG. 20. Mount Erebus on Ross Island, Antarctica. At 3,794 metres this is the second highest volcano in the Antarctic. It is a faint echo of the volcanoes that once punched through the icy crust of Snowball Earth.

They were lucky. The wind abated, the ice shower subsided and the next day—sunny and calm—Rothera Base airlifted a new tent to them. Two weeks later as they sledged back through their campsite on the return journey they looked on in horror as they realized that where they had camped was now a large depression sitting over a crevasse. If the wind hadn't got them, the crevasse would have swallowed them up: the legendary John Smellie had used up two of his nine Antarctic lives at one go.

We are lucky that John survived, for his work on the ancient volcanoes of James Ross Island shows that even during the warm early Pliocene the Antarctic Peninsula was glaciated, at least for some of the time.[9] On this evidence, the Peninsula land mass was, to some extent, insulated from the Pliocene warmth.

The Pliocene story, though, is more complicated. Between the thick volcanic lava piles of James Ross Island there are thin slivers of marine rocks, and rocks deposited by glaciers that have ripped up fragments of the ancient seabed around the island. This seabed was sometimes littered with molluscs and other organisms thriving in the seas this far south. These animals have, locked in their shells, their own story of Pliocene climate. It is one of greater warmth.

The shells of Cockburn Island

Cockburn Island rises out of the Weddell Sea, looking rather like a large flying saucer that has landed to the east of James Ross Island. It is visited most easily by helicopter from Seymour Island to the east, where the major Argentine Antarctic base of Marambio—with its supply of gourmet steaks—is sited. The precipitous cliffs of Cockburn Island fold out into a large flat-topped surface that you can reach, courtesy of a Royal Navy helicopter pilot. We are told that it is possible to climb up its sides, but at over 100 metres in height and covered in the guano of a major penguin colony that skirts its base, one would rather not. The wind-swept flat top of Cockburn Island rises to another small volcanic peak, but otherwise there is nothing to give one cover here, except for a small tin-roofed hut erected by Argentina, the only shelter from fierce Antarctic winds.

Just 3 kilometres long, Cockburn Island preserves the best fossil seashells from anywhere in the Antarctic. The rock deposit containing the shells is draped on volcanic rocks, but records a brief interval of climate in the early Pliocene that appears to have been warm. The seashells atop Cockburn Island have not seen the sea for more than four million years, but they are a kind of large pectenid (scallop), *Chlamys,* that you can still collect along the shores of the Americas and along the coast of Britain. These are large, fast-growing molluscs that might add a centimetre or more to their shell each year. It is this fast growth that makes them messengers of ancient climate.

The ancient *Chlamys* shells of Cockburn Island grew through the year, each of their tiny millimetre increments preserving a signal of perhaps a few days of weather. Their isotopic composition signals the temperatures of the waters they occupied. And, as the molluscs that built these shells appear to have 'lived fast' and grown rapidly throughout the year, they suggest that during warm intervals of the Pliocene there was little or no seasonal sea ice in the northern Weddell Sea. These molluscs, found as fossils in the rocks of Cockburn Island, do not resemble the molluscs clinging on for dear life around the coast of that island today. Rather, they are the close cousins of molluscs now living quietly in the coastal waters of Patagonia. *Chlamys* has retreated from the Antarctic, and the fossil record suggests this happened about three million years ago. What this retreat suggests is that once upon a time in the early Pliocene, the climate of the Antarctic Peninsula may have resembled Patagonia. It was much warmer than today, perhaps by as much as 5 or 6 °C in the summer months.[10]

Patterns of climate

Capturing a picture of Pliocene climate is not easy. It is, after all, a moving target, for climate within it—and even within the small segment of it that is being so exhaustively analysed by the PRISM research team—was continually changing. That change registers as the small-scale flickering seen, for example, in the seventy-million year Cenozoic climate curve painstakingly assembled by Jim Zachos and his colleagues from the evidence of the deep ocean oozes. When calibrated and statistically tested, this flickering emerges as essentially the same phenomenon that Grove Karl Gilbert had so presciently discerned in the Cretaceous strata of the USA interior, in those exciting and dangerous early days of geological exploration of what really was the Wild West. It is the signal of rhythmic, astronomically driven climate change—a signal dominated, in this part of Earth's history, by the 40,000-year change in the tilt of the Earth's axis.

This signature, viewed thus as if through the wrong end of a telescope, appears simply as modest oscillations in the Earth's temperature. What, though, is its real, physical meaning? The best places to appreciate the scale of such change is to get up close to where the change is expressed most vividly—and one such place is the Ross Ice Shelf of Antarctica. This is the Earth's largest ice shelf, covering nearly half a million square kilometres—twice the area of Britain. For some years, there have been concerns expressed as to how stable—or how unstable—it might prove to be in our warming climate. It is useful, therefore, to gain some understanding of how it behaved when the Earth was just a little warmer than today, in the Pliocene.

Gaining such understanding, though, is easier said than done. This kind of history *is* preserved—but it lies under the present ice shelf itself, within the layers of sediment that exist deep beneath the sea floor. To get at this history was expensive, and technically challenging. In something called the ANDRILL (Antarctic Drilling) project, a large and sophisticated drill rig had to be hauled out on to the ice, and then used to penetrate the ice, the icy sea water below, and finally the sea floor, from beneath which cores of sediment were brought up. The story was worth the effort.[11]

The Pliocene strata, representing some three million years, are about half a kilometre thick. This means that the sea floor was subsiding, and that the history that was preserved was near-complete and packed with detail; and, satisfyingly, this detail was dramatically variable. Figuring largely among the diverse types of strata were, first, proper, thoroughgoing glacial tills—there were quite a number of layers of these. When they formed there the overlying ice shelf was thick—thick enough to extend right to the sea floor and carpet it with ice-dragged debris. At the opposite end of the spectrum, there were interleaved layers of fine-grained oozes made up largely of billions of the tiny silica skeletons of diatoms, forms of single-celled planktonic algae. As these oozes accumulated on the sea floor, the ice shelf had collapsed or shrunk back and there was open water on the Ross Sea, on which drifted some icebergs, from which scattered blocks and pebbles could drop into the oozes. Intermediate conditions are represented by gravelly sediments, brought in by more continuous floating ice.

Overall, the picture is one of great dynamism, with the ice shelf forming and collapsing repeatedly in its history—and the pattern of the interlayering of the strata suggests that the control here—as with the deep-sea oozes—is the astronomical pacemaker. The pattern isn't quite metronomic—although the 40,000-year signal is dominant. There are more continuous clusters of ice-laid deposits, and also some thicker layers of the ooze, each representing several cycles in which the ice shelf did not return. This repeating construction and destruction of the ice shelf probably caused sea levels to rise and fall by some metres: not directly in the case of floating ice (because that simply displaces its own volume of water)—but because an ice shelf helps pin back ice in the interior. Once it is removed, the inland ice can flow to the sea that much more quickly.

For all the insights that they has provided, the ANDRILL cores gave no clue to the scale of sea level change. For that, one needs to travel far from the ice.

Pliocene high seas

The volcanoes and molluscs of James Ross Island can give clues to the Antarctic Peninsula ice sheet and the climate of the Weddell Sea in the Pliocene, and the sediments below the Ross Ice Shelf betray the history of the ice above. What, though, of the much larger West and East Antarctic Ice Sheets? These ice sheets have a large albedo effect, cooling the Earth by reflecting heat back out into space, and they also regulate the Earth's sea level, locking up between them over 90 per cent of all fresh water on the surface of the Earth. That is a *lot* of water, equating to about 65 metres of sea level rise if the entire Antarctic ice sheet were to melt, plus another 6 metres if Greenland melts too.

The Pliocene, in general, shows what the sea level was like the last time carbon dioxide reached 400 ppmv in the atmosphere. There are a number of ways of estimating changes in Pliocene sea level, including geography, changes in the isotopic composition of the shells of deep sea foraminifera (that reflect the balance of water between the oceans and the ice sheets), evidence from the Antarctic continent itself, and computer models.

Perhaps the most striking evidence for Pliocene sea level is the inland 'escarpments' that are a distant memory of ancient, higher seas. They are former sea cliffs, cut when the sea level was higher than today and now left stranded far inland. One of these scarps runs along the Atlantic coastal plain of the United States from Florida to North Carolina, called the Orangeburg Scarp after the town of that name in South Carolina. You might be forgiven for missing the scarp as you drive by, for after three million years of erosion it is just a gentle step in the landscape. But travel north-east on State route 33 from Orangeburg and you'll pass the Bull Swamp Cemetery, and turning left onto Till Road, you more or less parallel Interstate 26. Here the land gently rises from less than 60 metres altitude to a dissected upland surface that is just over 90 metres high. The Orangeburg Scarp was cut by the action of waves, and those waves extended tens of kilometres inland, and signal a sea level perhaps tens of metres higher than at present.[12]

FIG. 21. The Orangeburg Scarp extends from Florida across Georgia to North Carolina and represents the eroded and degraded cliff of an ancient Pliocene shoreline. On this satellite image it is traced as a curving line that mirrors the present coast.

We cannot be precise about the amount of sea level change. While the seas rise and fall as the ice sheets wax and wane, the land also rises and falls according to its plate tectonic setting. So the Orangeburg Scarp, like all other geographical features, contains a component of local (tectonic) and global (sea level) change. But the shells of marine foraminifera and ostracods, living on the seabed in the Pliocene, also provide an estimate of sea level. Their calcite shells have a record of the stable oxygen isotopes from the sea water in which they lived. And when ice sheets failed, and the sea level rose, light ^{16}O flooded into the world's oceans from the melting ice. As we have seen, that rise in ^{16}O is recorded in their minuscule shells, and it can be used to estimate the degree of sea level rise. Like the Orangeburg Scarp it also suggests sea levels higher than today's by more than 10 metres. At this time the sea covered wide areas of Florida. The quarries at Sarasota, on the west coast of Florida extract limestone for local roads. They are crowded with marvellous fossil animals: gigantic snails, the teeth of a multitude of sharks, and even the bones of whales. They are also a reminder, of course, that the sea may come in again as the climate warms.

Where did the extra ocean water in the warm intervals of the Pliocene come from? The West Antarctic Ice sheet clearly has not been stable, and melting it would add perhaps 5 metres to sea level. The Greenland ice sheet was probably much smaller in the Pliocene, at least until the Central American Isthmus closed, and this would add a few more metres to the sea level. So perhaps 10 metres of sea level rise can be found quite readily from the evidence to hand. But to raise the sea level by 35 metres, as some of the estimates suggest, requires part of the East Antarctic Ice Sheet to have melted. Yet the marvellous lake fossils from Mount Boreas (see Ch. 5) suggest that the East Antarctic Ice Sheet may, at the very least, have been stable for fourteen million years. Pinning down the extent of Pliocene sea level rise is still (like much else) work in progress.

Along the Armadillo highway

Spare a little thought for those little short-legged armoured creatures that cross your path on the side roads and forgotten highways of the southern States of the USA. On the small roads that circumvent the city of Tishomingo in southern Oklahoma you may have passed their little bodies stacked up beside the road, four legs pointing skywards and gradually picked clean by carrion birds. But the peculiar armadillos signal larger events in the late Pliocene. That they are here in North America at all is a gift of plate tectonics.

The ancestors of the Oklahoman armadillos lived in South America (on that Pliocene drive along the Pan-American Highway you would not have crossed their path until you reached Venezuela). The present-day victims of road-kill were American adventurers long before Cortes or Pizzaro. Armadillos, like many animals, came north when the Americas were finally united about three million years ago. The land bridge thus formed allowed the Great American Interchange. Bears, dogs, and cats headed south; porcupines, sloths, and the armadillos headed north.

The forging of this land bridge had a great impact on life on land. Its impact on climate may have been yet greater. About this time there is evidence for stronger growth of the Arctic ice sheet. Low latitude surface waters that once flowed from ocean to ocean between the two Americas were now deflected north, the warm waters extending to the Arctic, where they released their moisture as snow. That snow was to fall ever more thickly, to cover the land. The Central American Isthmus was not the only ocean gateway to close. The story of the Cenozoic has been one of high-latitude gateways opening and low-latitude gateways closing, with profound impact on polar climate. As we have seen, the Indonesian Gateway was also narrowing at this time, gradually strangling the throughflow of warm surface waters from the Pacific.

There is something to think about in the complexity of this beautiful little world in which we live, of the intricate and sometimes intimate interrelationships of the Earth System. If the tectonic plate that carries Australia and New Guinea had not broken away from Gondwana, drifted north, and gradually closed the Indonesian Gateway to warm Pacific water, then the Indian Ocean would not have cooled, East Africa would not have become drier, and the forests of that continent would not have retreated. The ancient primates that hopped and bounced among the branches of those Miocene and Pliocene forests would not have walked out onto an expanding savannah. They would not have looked to the far horizon to scan for predators, they would not have stood on two legs freeing their hands to pick up and manipulate tools. Had New Guinea not drifted north and closed the Indonesian Gateway, you would not be reading these words now.

The Pliocene warmth was not to last. The cooling world at the end of this epoch can be read in the deep-sea oozes of the north Atlantic and Pacific, and in signs that icebergs drifted into ever-lower latitudes, shedding rocky debris as they went. The fully glacial world is about to be born.

Into the Icehouse

Ymir's legacy

The frozen lands of the north are an unforgiving place for humans to live. The Inuit view of the cosmos is that it is ruled by no one, with no gods to create wind and sun and ice, or to provide punishment or forgiveness, or to act as Earth Mother or Father. Amid those harsh landscapes, belief is superfluous, and only fear can be relied on as a guide.

How could such a world begin, and end? In Nordic mythology, in ancient times there used to be a yet greater kingdom of ice, ruled by the ice giant, Ymir Aurgelmir. To make a world fit for humans, Ymir was killed by three brothers—Odin, Vilje, and Ve. The blood of the dying giant drowned his own children, and formed the seas, while the body of the dead giant became the land. To keep out other ice giants that yet lived in the far north, Odin and his brothers made a wall out of Ymir's eyebrows. One may see, fancifully, those eyebrows still, in the form of the massive, curved lines of morainic hills that run across Sweden and Finland.

We now have a popular image of Ymir's domain—the past 'Ice Age'—as snowy landscapes of a recent past, populated by mammoths and woolly rhinos and fur-clad humans (who would have been beginning to create such legends to explain the precarious world on which they lived). This image, as we have seen, represents a peculiarly northern perspective. The current ice age is geologically ancient, for the bulk of the world's land-ice had already grown to cover almost all Antarctica, more than thirty million years ago.

Nevertheless, a mere two and a half million years ago, there was a significant transition in Earth history—an intensification of the Earth's icehouse state that spread more or less permanent ice widely across the northern polar regions of the world. This intensification—via those fiendishly complex teleconnections that characterize the Earth system—changed the face of the entire globe.

Global signals of northern ice

The changes can be detected in the sedimentary strata that were then being deposited around the world. The expansion of northern ice cannot be seen, paradoxically, in those places where it grew—in North America, Greenland, Scandinavia. This is because the subsequent (and more extensive) ice sheets that grew there later, especially over the last million years, have scoured away virtually all evidence of that first significant growth of northern ice. One can see the results of this recent ice-scouring as one walks among the mountains of the Alps, or of Scotland, or the Rockies. Those classic glacial landscapes bear the traces of destruction, not construction. They clearly show the sculpting of the mountains by the last advances of ice—but that ice, as it ground its way to the lowlands, swept away the sediment laid down during the many—perhaps fifty or more—earlier advances of ice.

To find traces of the early spread of northern ice, one must go further—and deeper—to where ice-affected sedimentary strata from that time are still preserved. One good place to go is the bottom of the north Atlantic, where fine-grained sedimentary oozes have slowly accumulated for millions of years, far out of reach of the erosive power of moving glaciers. These deep-sea oozes provide a marvellous, continuous record of the evolution of global climate that continues to be investigated, in ever finer detail, by the Ocean Drilling Program. Much of the information can be read only with the help of high technology: existing, for example, in the form of subtle chemical changes to the shells of planktonic creatures (see Ch. 4). However, the growth of northern ice produced a signal that is clear to the naked eye.

As northern ice spread southwards, so more icebergs broke away and drifted into ever more southerly waters. These icebergs melted, and released their freight of land-derived cobbles and pebbles and sand—and these rocky fragments then fell to the bottom of the ocean. Borehole cores pulled up from the floor of the Atlantic Ocean show that icebergs intermittently drifted across the far north from more than ten million years ago. A little over two and a half million years ago, though, glacier-strewn debris suddenly spread much further south than before—a sign of rapidly expanding ice.

A northern chill spread far beyond the reach of ice. In Western Europe, forests dominated by warm tree communities begin to show significant phases of pine growth. In equatorial Africa, between three and two and a half million years ago, there was widespread drying, and replacement of forest by savannah; this was a major factor in the appearance, and subsequent evolution, of a then obscure mammalian genus, *Homo*. In China, windblown, silt-rich dust—loess—began to spread over the landscape, derived from the glacial erosion of the Himalayas. Overall, these changes represent the most fundamental rearrangement of the Earth system since the beginning of ice growth on Antarctica, at the beginning of the Oligocene Epoch.

These changes, too, have formed the basis of a geopolitical battleground, fought over the last few years. This battle—between

Earth scientists rather than armed troops, fortunately—was over the definition of a geological time term: the Quaternary. This term has been long regarded as representing the 'true' bipolar Ice Age, and yet its definition has been highly problematic. Its formal beginning (or more precisely, that of the first Epoch within it, the Pleistocene) was established more than half a century ago, based upon palaeontological changes in the Mediterranean—but at about 1.8 million years ago, rather than 2.6 million years ago, and this was subsequently formally ratified in 1985. Part of the geological community were happy with this definition, and wished to retain it—not least to preserve the stability of the Geological Time Scale. But, most scientists working on Ice Age phenomena were unhappy with the situation, as it became increasingly clear that the formal definition did not match the 'natural' boundary within Earth history.

There followed the most almighty kerfuffle. The atmosphere between the opposing camps became heated—even vituperative. Words were exchanged, most pointedly. It may seem strange to a disinterested observer that so much passion can be expended on a mere technical definition—but words and their precise meanings are important, especially to scientists. In the end, the majority won the day, and the Quaternary Period is now defined to begin at a little under 2.6 million years ago.[1] Many of those involved (we too) still bear the scars, though this definition seems now set to stay.

At the beginning of the Quaternary, hence, the world changed, not in some form of isolated revolution, but as a very significant step within a complex, long-term cooling trend. But what caused the transition to a fully glacial Earth? There are many clues—a bewildering number of them, in fact. Among the contenders for this particular step change, mountain belts, ocean gateways, and particular combinations of seasons and ocean currents have received considerable attention.

Mountain masses and ocean gateways

Mountain belts, as we have seen, can affect climate by acting as a kind of complicated chemical sponge for carbon dioxide, and this (Ch. 5) may have been a factor in bringing in the Cenozoic glaciation. But they can act in other ways too. The simple bulk of a newly elevated mass at the Earth's surface can alter global wind patterns—and consequently ocean current patterns, too, as these are partly driven by winds. A newly risen part of a mountain belt might therefore set up conditions more conducive to the growth of polar ice—for example (it has been suggested) by driving warm currents further into the north Atlantic, to act as a moisture source to feed snowfall to help the North American and Greenland ice sheets grow. However, the precise amount and timing of uplift remain poorly established (for mountain belts, by their very nature, are erosional areas, so direct evidence of their height in the ancient past is difficult to establish). Furthermore, computer modelling of the effects of such mountain uplift suggest that, in itself, this mechanism is too small to drive ice growth in the northern hemisphere.

Some connection of glaciation and mountain belts is suspected. Many mountain belts around the world, in the last few millions of years, in general seem to have been rising more quickly. This strangely coincidental rejuvenation was based on indicators of the height of mountains in the past compared to the present, such as fossil plant assemblages; and on evidence as to how rapidly they were being eroded, notably the amount of sediment accumulating as strata around them. If there really had been a global phase of increased mountain growth, then that might in turn have removed atmospheric carbon dioxide as more rock was exposed to be weathered, thus paving the way for glaciation. But, there seems to be no sensible mechanism to produce such a suddenly more dynamic Earth.

However, mountain belts may be eroded more quickly (and hence also may rise more quickly, by the phenomenon of isostatic rebound) *because* glaciers are growing on them and wearing away the rocks beneath. Then, cause and effect may be reversed, in a classic example of a 'chicken and egg' situation.

So, if rising and eroding mountain belts are a mechanism for global glaciation, quite how this works is still so complex and imperfectly understood that most scientists put it in a file marked 'pending' (but under continuing study).[2] However, there are climate-altering mechanisms that take place on lower ground. The overall pattern of land and sea is continually, slowly, changing, and some of the new geographic arrangements might, literally, have provided a gateway to a global glaciation.

Something over five million years ago, the sea floor between north and south America—which had, for many millions of years been separate continents—began to rise. The sea between the two continental masses began to shoal. A little more than two and a half million years ago, as we have seen in the previous chapter, there appeared a narrow strip of dry land linking the two continental masses.[3] The Central American Isthmus had formed, and with that there came consequences. There was, as we have seen, that mixing of biological communities, of giant sloth and armadillo and terror-bird, as a corridor formed between long-separated ecosystems. But, the effects may have reached much further.

When the Straits of Panama were open, there would have been mixing between Pacific and Atlantic sea water, making their densities and salinities more similar to each other than they are today. This, it is surmised, would have slowed the 'global conveyor' of deep ocean currents, and in particular would have weakened the Gulf Stream in its journey from the Caribbean region towards Europe. This current not only heats north-west Europe—it also brings with it moisture-laden air that, turned to snow in the far north, nourishes the Greenland ice sheet and other ice-masses in the region. When the Straits finally closed, and the Americas joined, the changing of the Atlantic circulation pattern to a 'modern' one has been seen as a key factor (by some, *the* key factor) in promoting the massive growth of ice in the northern hemisphere.

Again, this neat idea—or at least any nicely pat and simple version of it—is, when closely examined, not a complete answer. For a start, the isthmus took a long time to rise before breaking the surface, and would have affected currents well before the final 'land bridge' state was arrived at. That might even support the idea to some degree, for the north began cooling, and significant ice-rafted debris began to fall on to the Atlantic sea floor, from more than five million years ago. But it makes it harder to implicate clearly the Central American Isthmus in the major climatic step at the beginning of the Quaternary, at some 2.6 million years ago—and, furthermore, as with the rising of mountain masses, computer models suggest that the formation of this land bridge would, in itself, have been insufficient to bring about a full, bipolar Ice Age.

Nevertheless, major geographic changes like this could not have happened without *some* consequences for the Earth's climate, given how important ocean currents are to distributing heat around this planet, and how sensitive they are to being reorganized by relatively small, tectonically induced changes to the shape of the ocean floor. The Panama Isthmus is not the only potential culprit. We have seen (Ch. 6) how, a little earlier, some three to four million years ago, severe constriction had begun to develop in another gateway, the Indonesian seaway, as New Guinea and Australia inexorably moved northwards towards Borneo. This would have slowed and then stopped the flow of warm South Pacific waters past Indonesia, replacing this flow with colder, North Pacific waters. The Indian Ocean, consequently, cooled slightly, and this change has been implicated (by further modelled teleconnections) in the drying of central Africa—and in less heat being transported from the equatorial to the polar regions, to encourage the growth of polar ice.

It is a bewildering array of possibilities. And there are more, too. Other mechanisms have been suggested, and most have largely fallen by the wayside for want of supporting evidence. It has been suggested that the Earth went through a particularly dust-rich part of space, the dust then absorbing some of the Sun's earthbound light and heat. And a particularly fierce bout of explosive volcanism worldwide has been proposed as a mechanism to throw up enough fine ash and sulphur particulates to cool the atmosphere and encourage ice growth.[4] These are among the also-rans—but there are other, more realistic influences.

One factor is simply astronomical, changes in the Earth's spin and orbital patterns, now known as Milankovitch cyclicity (see Ch. 4). Between three and two and a half million years ago there was a particularly favourable astronomical pattern that made for—on the whole—slightly cooler summers, allowing snow to persist a little longer through the year. Given how important the Milankovitch mechanism has been shown to be as a driver of climate in both non-glacial (Ch. 4) and glacial (Ch. 8) times, it seems reasonable to include astronomy as a prime suspect—at least as accomplice—in the refrigeration of the Earth, and in the birth of the Quaternary landscape.

And there is another part of the jigsaw: carbon dioxide. This greenhouse gas is clearly implicated in climate change, and looms large as another key suspect. We know that typical Quaternary levels—at least until humans began to industrialize the planet—varied between about 180 ppmv (in cold phases) and 280 ppmv (in warmer phases) in the atmosphere. Directly measuring the carbon dioxide content in pre-Quaternary times is difficult, but we have seen (Ch. 6) that the atmospheric concentration in the Pliocene Epoch, the last gasp of relative global warmth, was likely to have been about 400 ppmv. If this was so, a significant reduction in carbon dioxide levels, whether by increased rock weathering or by some other mechanism, must have taken place around the transition into the Quaternary icehouse.

But among these and other competing and concatenating factors, there is one pattern that has recently emerged that forms, if not a final answer, at least a plausible and sophisticated means of placing large amounts of ice onto northern hemisphere land masses: it is the Pacific snowgun.[5]

A snowgun in the Pacific

On the floor of the north Pacific Ocean, south and east of the spectacular, volcano-studded Kamchatka Peninsula, deep sea oozes have been slowly accumulating for millions of years—as is typical of ocean floor settings throughout the world. These particular sediments, though, in layers that are dated to be a little more than two and a half million years old, show a curious, and seemingly contradictory, pattern of characters. Yet, they might represent a key part of the Ice Age puzzle.

For a start, the sedimentary layers at this level show, as expected at this time, an increase in the amount of ice-rafted debris, indicating that, as in the north Atlantic at this time, northern ice was beginning to push southwards at the start of the Quaternary Ice Age. Yet, these sedimentary layers also held within them various microfossils, and these send seemingly mixed messages about the state of this particular part of the ocean.

The oxygen isotope patterns in the calcium carbonate skeletons of fossilized planktonic single-celled micro-organisms, the foraminifera (see Ch. 4), do indeed suggest cooling of the sea surface, in agreement with the spread of icebergs. Yet, fossilized coccoliths—parts of the calcareous skeleton of microscopic algae called coccolithophores—suggest that the ocean surface here, rather, was *warming* instead of cooling. The evidence here is from a set of remarkably useful chemicals—long-chain alkenones—produced by these algae, the particular composition of which, in modern forms, is directly related to temperature. These alkenones can be extracted from ancient sediments and analysed and—assuming that ancient coccolithophores had similar biochemistry—can be used as a direct thermometer of sea surface temperature in the geological past. The signal from these Pacific sediments shows that, as the icebergs massed above, temperatures rose by as much as 7 °C. Are the results reliable? The alkenone geothermometer in other studies has given what seem to be generally trustworthy results. In the case of these north Pacific ocean floor sediments, too, the mysterious warming has also been shown by the chemistry of the skeletons of another group of microfossils, the silica-shelled diatoms. At the same time as these contradictory patterns were being established, yet another change was taking place—the general abundance of the diatoms was dropping precipitously, because they were being starved of nutrients.

How could these diverse pieces of evidence be put together? First, by considering the biology of the different groups of fossils. The coccolithophores and diatoms bloom in summer, while the foraminifera prefer the winter and spring. And therefore, the pattern is one of colder winters and warmer summers—and thus a change in seasonality. But how did the waters heat up so much in the summer? One possibility is that there was less mixing with the cold underlying waters. A powerful indication that the mixing had indeed been steeply reduced is that the deeper waters are also nutrient-rich—and the marked drop in the abundance of the diatoms shows that the nutrient supply from below had been cut back sharply.

With warm waters persisting long into the summer, more moisture-laden air can then rise into the atmosphere and be blown across the North American continent. And, in the colder winters, it will fall as snow—and begin to stick, and the great northern ice sheets will begin to grow. In this way a gigantic snowgun is created and brought to bear on an entire continent.

The snowgun idea is attractive, not least because of the vivid analogy of its name. The idea is plausible—perhaps even persuasive (and something like it might, just perhaps, have driven the Ordovician icehouse as we noted in Ch. 3). There is, of course, the question of *why* the north Pacific became more stratified at this time, and this is, so far, rather more generally ascribed to wider-scale climate changes associated with the Earth's orbital characters and with carbon dioxide levels. No matter, the evidence is mounting: the sequence of events around the inception of the bipolar Ice Age is being ever more closely described and analysed. Perhaps we are nearer to solving the mystery of why the bipolar Ice Age of the Quaternary started, *how* it did, and *when* it did.

Now that we have entered the Quaternary Ice Age, though, it is time to examine its extraordinary history. It represents, perhaps, the epitome of the Earth machine in all its finely balanced complexity.

The Glacial World

Beginnings

It is a scene of devastation, as far as the eye can see. Swathes of bleak landscape, with strewn boulders embedded in a sticky mass of sandy clay. Here and there are signs of a little more order—distinct spreads of gravel or patches of fine sand. Mostly, though, it looks as though every type of sediment, from fine clay to house-sized blocks, has simply been stirred together and spread across the land. Remove the crops and topsoil of gentle Leicestershire and Suffolk, or of central Germany or Kansas, and this is what lies beneath. Between the ordered sedimentary strata of the distant geological past and the ordered calm of the present is evidence of an only-just-elapsed catastrophe, and two centuries ago, when the science of the Earth was young, the naturalists of those days pondered on what it might mean.[1]

There were those like the young William Buckland, both Reader in mineralogy at Oxford and priest (he went on to become Dean of Westminster), who saw in it evidence of the biblical Deluge. Or Jean André de Luc, mentor to the wife of George III, who considered that the large blocks had been fired, like Roman ballista, from the mountains by some powerful but mysterious explosions. Or Sir James Hall, a savant of Edinburgh, who thought that the blocks had been carried into position by tsunamis, generated when large areas of sea floor (he supposed) suddenly popped up like blisters—he was clearly of an intellectually playful disposition. Or Leopold von Buch, who invoked catastrophic mudflows (one such, indeed, did take place in an Alpine valley, the Val de Bagnes, just after von Buch's paper on this topic was published, when a natural dam burst, scattering mud and boulders far down the valley, and killing many people). But it was that extraordinary polymath, Johann Wolfgang von Goethe (a one-time Superintendent of Mines, if you please) who was among the first to sense what had been going on, when he associated the scattered blocks with a great expansion of the Alpine glaciers he was familiar with, and coined the term *Eiszeit*—the Ice Age.

It was a controversial idea, and remained so for decades. Among those who lived amid—and sensed the power of—the ice of the Alpine mountains, men such as Jean-Pierre Perraudin, Ignatz Venetz, and Jean de Charpentier in the early nineteenth century, it made sense. But for most people the idea that a sea of ice had spread across the calm European landscapes in the geologically recent past was simply unbelievable, outrageous. It took the energy, persuasiveness, and prestige of one of Europe's leading scientists, Louis Agassiz (who travelled to the Alps to demolish the upstart glacial theory, and came away a convert) to argue the glacial case and—eventually—to win over the scientific establishment, including the larger-than-life William Buckland, a man omnivorous both in the intellectual and the gustatory sense (he ate his way, it is said, through the entire animal kingdom).

Towards the end of the nineteenth century, the idea of a world submerged under ice had taken hold. That idea, though, quickly became more complex. Firstly, it was not an entirely ice-covered Earth—not a global snowball, as was to be envisaged by later generations of scientists for much earlier times in Earth history (see Ch. 2). It quickly became clear that the stuff known as boulder clay—that chaotic mixture of clay and sand and pebbles and scratched and grooved boulders (now called, more technically and less poetically, glacial till)—had largely been smeared on to the landscape from the underside of a moving mass of ice. Tracing out the distribution of this material, it was seen that the ice sheets, at their maximum, had found their way, for example, into north London—but not into the southern part of that city. Beyond the ice limit (in what is now south London, for instance), there was evidence of bitterly cold periglacial conditions—masses of gravel deposited by seasonal melt waters, with the ground in between cracking from the cold. In this frozen ground there grew wedges and blisters of ice, as in Siberia today, while it was also swathed in wind-blown silt known as loess, blown across the land by the strong glacial winds. Yet further south, these indicators of adjacent and nearby ice died away, to be replaced by sediments typical of more temperate conditions.

It soon became established that the Ice Age was not singular, an uninterrupted epoch of unremitting cold, for here and there, as interlayers within the masses of glacial debris, were sediments with evidence of warmer conditions—fossil remnants of pine and hazelnut and oak, and bones of deer and wood mouse and hippopotamus. It was clear that climate within the Ice Age was prone to dramatic change. But how many times had it changed like this, alternating between icy and warm conditions?

This was difficult to establish by the classical means of land-based geological study. Geological history is typically read by establishing successions of strata, tracing these across country, and interpreting the clues to climate and environment preserved within them. But on land, the Ice Age strata are irregular, discontinuous, fragmentary, more often than not cut through by erosion, or partly bulldozed away by advancing ice sheets. Furthermore, it was extraordinarily difficult to establish their age precisely enough to establish an ordered sequence of events. The glacial deposits generally lacked fossils that dated from the time that they were formed, partly because little lived in those inhospitable landscapes, and partly because those fossils that they did contain had often been excavated and transported from older strata. Moreover, the glacial strata left from successive ice advances tend to resemble each other, simply because as ice advances anew each time, spreading debris from the mountains across the plains beneath, it is inclined to follow much the same path—and therefore successive ice advances leave similar-looking deposits of glacial till and sand and gravel.

The sedimentary layers that commonly contain trustworthy fossils are from the warm interglacial phases. Such strata, though, typically only occur as irregular pods and lenses, representing, say, the infills of ponds and old river channels. It was generally impossible to physically track these across country, to see how they fitted into each other or into other Ice Age successions elsewhere. To add to the difficulties, the fossils within them—that could often be abundant—closely resembled each other, because the time elapsed between successive warm phases was so short that the animals and plants showed little evolutionary change from one phase to another.

Distinguishing between the deposits left behind by successive warm phases depends on carefully charting and comparing the order and abundance in which species appeared and disappeared as the climate warmed and then cooled. One warm phase may be distinguished by a notable abundance of elm, say, or another by the particularly early appearance of beech or of hazelnut. Looking at the animals, one might distinguish, for instance, a warm phase where wild horses spread into and colonized Britain (only to die out when the ice pushed down again), versus another where horses were absent from this island, because the sea level in the English Channel rose too quickly for them to recolonize from Europe as the climate warmed.

This is slow, painstaking detective work. Even so, it eventually led to a consensus by about the mid-twentieth century that, across much of the northern hemisphere at least, there had been around four major advances of ice, separated by warmer phases, the last of these being the one we currently live in. And, before the first major incursion of ice, evidence was being found of, perhaps, a roughly similar number of oscillations between cold (but not quite glacial) and warm climate in the older part of the Quaternary Period.

This was, though, a seriously misleading consensus, not least because it obscured the *causes* of climate change. The idea of astronomical control of Earth's climate, of those variations in Earth's orbit and tilt and spin, had been proposed in the late nineteenth and early twentieth century by the likes of Joseph Adhemar, Grove Karl Gilbert, and James Croll, and then meticulously elaborated by Milutin Milankovitch (see Ch. 5). But it seemed to be quite inconsistent with the pattern of cooling and warming events inferred from the historical pattern gleaned from the land. There seemed to be simply too few climate changes to fit with a model of astronomical control of climate. So the 'Milankovitch hypothesis' (as it came to be known) was generally held to be an ingenious idea, but one that sadly did not fit the available evidence.

Astronomy and oceans

It was that most characterful of scientists, Cesare Emiliani, who we met in Chapter 5, who suspected that Milankovitch and his spin- and orbit-calculating predecessors might have been on to something. In analysing, in the 1960s, the oxygen isotope ratios of the calcium carbonate shells of fossil planktonic foraminiferans entombed within the beautifully continuous ocean floor sediments, he recognized a repeating isotopic pattern of higher and lower values. He interpreted these as representing alternating glacial and interglacial phases—and that there were more of these than the land record had suggested. Emiliani wondered if the Milankovitch hypothesis might, after all, be correct. But, he could not prove it, because he could not place the events that he recognized into a well-constrained framework of geological time—and so could not demonstrate whether they really fitted the astronomical prediction or not.

It was Nick Shackleton who added the time dimension. That multifaceted gentleman of British science had proven Emiliani wrong in interpreting the oxygen isotope signal in foraminifera as dominantly temperature-controlled (he showed that global ice volume mainly drove the isotope pattern). Now, he showed that Emiliani's suspicions regarding the origin of the glacial–interglacial cycles were essentially correct: that is, that they arose from the astronomical variations that Milankovitch had calculated.

Emiliani had lacked a well-founded timescale within which to place the isotopic patterns that he had discovered. Shackleton, together with the American researchers John Hays and John Imbrie, obtained a continuous set of isotopic patterns in a long, deep-sea borehole taken from the western Pacific Ocean and published in 1976 (formally labelled as V28–238, it has more vividly been called the Rosetta Stone of the Ice Ages). This borehole core went down from the sea floor to penetrate an unbroken succession of strata that included a major magnetic reversal (when the south pole became the north pole, and vice versa) which had been traced on to land, where radiometric dating had shown it to be 700,000 years old. At last—here was a robust time framework. Using this, Hays, Imbrie, and Shackleton, in 1976, published evidence that clearly showed the pattern of these climate changes and unambiguously fitted the astronomical pattern so painstakingly calculated—without the benefit of computers—by Milutin Milankovitch and his predecessors.[2] The strata in borehole V28–238 showed major 'pulse-beats' of 20,000, 40,000, and 100,000 years, representing the Earth's 'wobble', the tilt and stretching of its orbit around the Sun.

Was there complete control of climate by the pattern of the Sun's rays? Not quite, for the detail of the pattern showed significant differences—or rather modifications of the astronomical signal. Firstly, the astronomical signal is more or less symmetrical, with increases and decreases in incoming sunlight at any part of the globe being of generally equal durations. But, the observed pattern of climate mostly showed slow coolings and then abrupt, rapid warmings—a sawtooth curve. The changes in climate observed, too, particularly at mid-latitudes, from bitterly cold to pleasantly temperate, involving mean temperature changes of about 5 °C, seemed too great to be explained by the modest changes in sunlight brought about by the Milankovitch mechanism. Clearly, some aspect or aspects of the Earth system were acting to amplify and distort these small changes in sunlight. And, as more sediment cores from the ocean floor were examined, to give an ever longer and more detailed indication of the course of climate, the Quaternary Ice Age was found to show a distinct progression. However, it did not fit with any straightforward prediction based simply upon astronomy.

The two worlds of the Ice Age

In the modern world, the science of the Earth—as that of biology, chemistry, physics—is beset by data: measurements, analyses, observations that are extracted from the strata by thousands of scientists. These scientists have at their disposal the latest in analytical hardware, that can churn out analyses a hundred—or a thousand—times more quickly than could be got by earlier generations of laboratory equipment. It is easy to be overwhelmed by the flood of new data, to become hopelessly lost amid the embarrassment of scientific riches.

But, if one can manage to cope with it, to make use of this abundance of information, to piece it together—then sometimes a single, simple, sharp image can result: a new window through which to view the Earth's past.

A few years ago, the palaeoclimatologists Lorraine Lisiecki and Maureen Raymo, with infinite care, stitched together fifty-seven separate, partly overlapping segments of Pliocene and Quaternary climate history, as extracted from the very many cores of ocean sediment that had been analysed by different scientific teams from all around the world.[3] This climate history was shown in terms of what has become a kind of universal language, of mean annual sea surface temperature as worked out from the oxygen isotope ratios from the shells of fossil plankton, the foraminifera. It was a most painstaking task to splice these records together, like the assembly of a single music recording from dozens of individual fragments of tape recording, each made at a different time and sampling differing parts of the whole *œuvre*—and with some of the individual tapes having sporadic, invisible gaps where nothing was recorded.

The result, though, once published, quickly became well-nigh iconic. Lisiecki & Raymo (2005) shows, in one diagram, the course of Earth's climate over the past five million years. It is a pattern worth dwelling over. The main part of the diagram is a set of very many sharp ups and downs, like the seismograph of an earthquake. There is the pattern of the Milankovitch cycles of warm and cold climate—very many of them, and appearing so tightly compressed at this scale that it is hardly possible to tell one apart from the other. No matter—it is the overall pattern that is important, and that is what this particular compilation brings out so well. Towards the end of pre-Quaternary times, the dominant pattern is that of a steady 40,000-year pulse—the variations in tilt of the Earth's spin axis.

This is the '40,000-year world' and it continued through into the Quaternary, although at the transition into the bipolar glacial world there was a distinct downwards deflection in the pattern, as the world became colder. In the 40,000-year world, the difference between the cold and the warm phases of climate was distinct, but

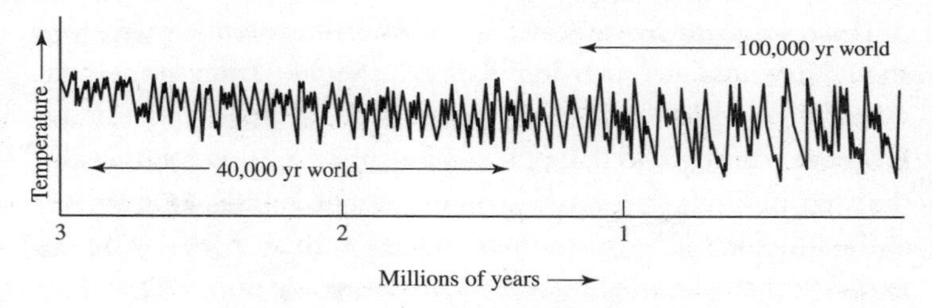

FIG. 22. The oscillating climate of the past three million years switched between glacial and interglacial, paced by changes in the Earth's position relative to the Sun, its wobble and its tilt (see also Fig. 13). Over half a million years ago the glacial–interglacial oscillation changed from a dominant periodicity of 40,000 years to one of 100,000 years, reflecting a change in dominance of eccentricity over obliquity. The reasons for this change are puzzling, but may relate to the changing stability of the ice sheets as these became more extensive.

still modest. Superimposed upon this pattern were episodes that were slightly warmer or distinctly cooler than their neighbours—and in the cooler episodes, say, permafrost could affect mid-latitude Europe—but ice sheets did not yet spread across the lowlands. Ice, though, may well have grown in mountainous areas, to swell the flow of lowland rivers during the spring melt.

Around a million years ago, the pattern changed, and so did the world. Some of the peaks became larger, with greater highs and lows, and these began to appear regularly, spaced about a hundred thousand years apart. This is the frequency of the variations in the Earth's orbit around the Sun, and it is the beginning of the '100,000-year world'. Now, during the great downward plunges of temperature, the ice grew beyond the mountains, and spread as enormous sheets over the lowland plains, to reach, at their maximum, London, Washington, and Berlin. Following each ice advance, there was a rapid thaw to bring about the warmest part of the cycle—the interglacial phase. It is the start of the true Ice Age, and also that of popular imagination. It is also a mystery.

There is no astronomical reason why the 100,000-year cycle should have attained such dominance. It should—from simple consideration of variation in sunlight received—be in general a minor influence compared to that of tilt and wobble. And, astronomically, the 100,000-year cycle was weakening, not becoming stronger, just at the time that it began to impress itself with such clarity on the Earth's climate. So what was—and still is—going on?

It remains a puzzle—though scientists have not been slow in coming up with ideas to explain this phenomenon.[4] It seems to hinge on some aspect of the Earth system evolving to select and amplify this particular one of the Milankovitch pulse-beats. For instance, it may be that as the ice sheets slowly grew larger through the Quaternary, they attained sufficient inertia to resonate only in sympathy with the longest of the Milankovitch cycles. As a variation of this, it has been suggested that, early in the Quaternary, the northern ice sheets grew out over land surfaces mainly covered by soft, loose sediment, over which they could both advance and decay quickly. Later, when much of this loose sedimentary cover had been scoured away, the ice sheets grew and receded over solid bedrock—and hence more slowly. These suggestions are best regarded as tentative; some of the workings of the Earth's climate remain profoundly enigmatic.

The message in the ice

The ocean drill-cores have provided us with an extraordinarily detailed picture of Quaternary climate history. But there is a limit to the sharpness with which they allow us to view this history, for nearly all the modern ocean floor has been well oxygenated for the past hundred million years or so. This means that the sedimentary layers have been thoroughly crawled over, burrowed through, *eaten* through by a variety of sea floor animals—worms, crustaceans, sea urchins, and yet others. Therefore, on a detailed level, the information within them has been shuffled and mixed beyond hope of recovery. This has not affected the broad outlines of the historical evidence that they contain—such as the major, multi-millennial Milankovitch cycles—but means that finer-scale historical information, dealing with the course of climate over years, decades, and centuries, has been obliterated almost everywhere on the sea floor. And, for all the extraordinary range of information that can be gleaned from the sediments and from the fossils buried within them, they remain tantalizing (and often downright infuriating) on some aspects that we would like to learn a *lot* more about, such as the composition of the Earth's ancient atmosphere.

There is, though, one stratal record that has not been chewed through by burrowing animals, and that has fossil air preserved within it. This is the history preserved within the ice layers of the great ice sheets of Greenland and Antarctica. It is a remarkable archive. Every year, in the interiors of these ice sheets, snow falls, and buries last year's snow—which in turn has buried the snowfall of the previous year, and so on. The eventual fate of this material is to flow to the edge of the ice sheet, and be released as glaciers. Yet, the history it preserves stretches deep into the Quaternary ice ages. In the centre of Antarctica there are places where some one million annual snowfalls are still left, stacked on top of each other (on Greenland, it is more like a hundred and thirty thousand annual snowfalls, for the snow piles up faster there—but the ice also flows more quickly into the sea).

The amount of different types of information in these snowfall-layers is prodigious.[5] The water molecules of the ice contain lighter and heavier isotopes of both oxygen and hydrogen, and different proportions of these can reveal the temperature of the air from which the snow crystallized. The snow, when it settles, is light and fluffy, full of air. As more snow falls on it, it is buried and gradually becomes compacted, and the air begins to be squeezed out of it. For a while, the air in the snowpack can still exchange freely with the atmosphere, the pores in the compacting snow remaining interconnected. With further compaction, though, the ever denser snow turns into solid ice and the interconnected air passages turn into isolated bubbles. It is at *this* point, several metres below the surface, that what is left of the air content is locked away in the ice. Thus, the air in the bubbles (from which levels of carbon dioxide and methane can be measured) is significantly younger—typically, by some hundreds of years—than the ice crystals that immediately enclose them.

There is more information, too, captured by the ice layers. They contain terrestrial dust that show that the glacial Earth was dry and dusty, while during interglacial phases it was wetter and less dusty. Rare, enormous explosive volcanic eruptions (such as Krakatoa and Pinatubo) can leave detectable levels of fine ash and sulphur. In the last couple of millennia, lead has entered the ice layers, first from smelting dating back to Roman times and more recently from the lead tetraethyl that used to be added to petrol. Yet more recently, industrial soot has made its way to the ice, and then the radioactive isotopes from atomic bombs.

This multifaceted history, though, needs to be extracted from the ice sheet—a daunting task. A drilling rig that is capable of penetrating, coring and recovering kilometre thicknesses of ice needs to be set up, manned, and operated in some of the most inhospitable conditions in the world, in the interior of the Antarctic or Greenland ice sheet. One of these remote drill sites is 'Dome C' in the Antarctic, over 500 kilometres from the nearest occupied research base, and over 3,000 metres high on the East Antarctic plateau. Here, even summer temperatures can dip below −40 °C and winter temperatures much colder than that have been recorded. Metre by metre, it took ten years to drill through 3 kilometres of ice at Dome C. It was time well spent. The ice core has revealed a fabulous record of past climate.

The climate history is magnificently detailed and extends back, currently, to a little under 800,000 years in Antarctica.[6] (As we write, drilling is ongoing to attempt to locate ice that is a million years old or more, to try to penetrate the 40,000-year world of the early Quaternary.) Individual annual snow layers can be recognized to depths of tens or hundreds of metres, equivalent to ages of centuries to millennia. Below that, the ice layers are generally too tightly squashed together to be separately recognizable, but, even so, the sampled units each represent only a few years of snowfall. The layers cannot be directly dated (even the volcanically derived layers—recognized typically by the high electrical conductivity of the ice, from its sulphur content—contain too little material to allow radiometric dating of the debris). But the pattern of glacial and interglacial intervals is clear, and can be directly and unambiguously correlated with the same events recognized in the ocean floor successions.

The ice cores, too, offer what the ocean floor deposits cannot provide—a comparison of temperature at the time the snow was falling (from the hydrogen and oxygen isotopes in the ice) with the composition of the atmosphere (from analysis of the air bubbles trapped in the ice). The two types of information, combined, show a close match: temperature and greenhouse gas levels rise and fall essentially in lockstep, right to the present (or, to be more precise, to about the year 1800, when the effects of the Industrial Revolution began to make themselves felt—but more of that anon).

At the heart of the match is a fundamental question—is the temperature rise and fall the result of almost instantaneous response of the climate system to changes in greenhouse gas levels? Or, are the changes in concentrations of carbon dioxide and methane the *result* of warming and cooling of the Earth's surface and oceans because of, for example, release of carbon dioxide from oceans as they warm and of methane from wetlands as they expand, with the climate itself being controlled by other factors, such as systematic variations in sunlight related to the Milankovitch mechanism?

This is a complex and much-discussed question. Let us try to tease apart some aspects of it, at least. Firstly, the Milankovitch

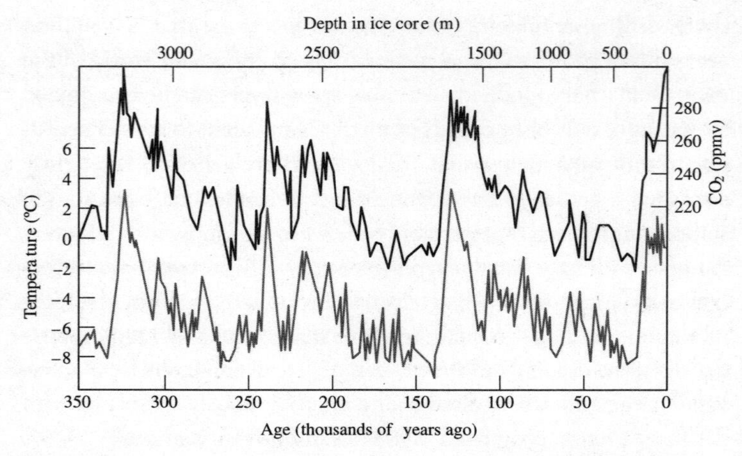

FIG. 23. A 350,000-year history of temperature and atmospheric carbon dioxide levels extracted from Antarctic ice.

signal, in both ice cores and ocean floor sediments, is clear. There is no doubt that seasonal variations in the Sun's energy are a crucial part of what is ultimately driving the glacial–interglacial alternations. However, as we have seen, the correspondence is not exact. The climate cycles, unlike the astronomical ones, are strongly asymmetrical—and also the astronomical cycles can be 'stretched' to fit other factors. In the last '100,000-year cycle', the time that separates us from the last full interglacial phase is in reality some 125,000 years—and in this case the 100,000-year cycle actually seems to be a multiple (in this case a fivefold multiple) of the 21,000-year precession or wobble cycles. So, the astronomical influence on climate is powerful, but subtle, because the different cycles of the Milankovitch mechanism are not simply additive but interact in complex fashion; other factors are involved, too, to amplify and distort the astronomical signal.

Secondly, there is the question of timing between changes (say, increases) in the carbon dioxide concentration and in temperature.

The record shows that temperature begins to rise slightly earlier than do the ice-bound carbon dioxide levels, following which both rise together. This might suggest that it is temperature that is driving the change in carbon dioxide. However, while warming ocean water will allow it to store less carbon dioxide overall, and cooling it will allow it to store more, this factor will only account for a small part of the change in carbon dioxide levels between glacial and interglacial phases. There remains uncertainty, too, in the precise relative timing of the temperature and atmospheric composition changes, because of the mismatch between the preservation of temperature signal in the ice and the time at which the fossil air is finally locked into the ice-enclosed bubbles.

Thirdly, there is no doubt that carbon dioxide is a heat-trapping greenhouse gas (as was so clearly shown by John Tyndall's table-top experiment (Ch. 1), or by the example of the planet Venus, kept at literally infernal temperatures by very high levels of this gas in its atmosphere). There remains uncertainty as to exactly how sensitive the climate is to carbon dioxide, relative to other influences (estimates vary by a factor of about two). It is most reasonable, though, to suggest that higher levels of this gas will cause warming, even if this warming itself, by means of positive feedback, will further increase its concentration in the atmosphere. It is most likely—and is the mainstream opinion in science—that astronomical changes stimulate initial warming and act to increase carbon dioxide levels, and that increase, then, strongly amplifies the original warming.

The case of the missing carbon

Given its staggering complexity, it is not surprising that the workings of the world should be so mysterious and difficult to fathom. But perhaps more staggering, as revealed by that magnificent ice core archive, is that its control of carbon dioxide is so finely managed.

Over the last 800,000 years levels of carbon dioxide in the atmosphere have varied regularly and systematically, almost metronomically, ranging between about 180 ppmv during glacial maxima and 280 ppmv during interglacials. This is managed, too, in a world where the carbon stores are huge: vegetation on land contains about 600 billion tons of carbon, roughly the same as the atmosphere; soils a further 1,500 billion tons or so. The oceans continuously exchange carbon dioxide with the atmosphere, contain over 35,000 billion tons—and that is not counting the stores in sea floor sediments, and in the sub-seafloor groundwater that courses very slowly through these sediments, and even through the basaltic bedrocks that lie beneath.[7] One might, intuitively, have thought that carbon dioxide levels should oscillate considerably more irregularly, especially given the sheer scale of environmental change during the Ice Age—let alone the effects of, say, occasional volcanic super-eruptions. The delicate pattern observed implies a formidable control mechanism possessed by the Earth.

What, though, was that control mechanism? Or, to take a step backwards, as interglacial gave way to glacial, where did about a third of the content of atmospheric carbon (about 200 billion tons) disappear to—subsequently to re-emerge at the next warming? This has been another of the great puzzles of the Ice Age, and, while the circle of suspects seems to be narrowing, it would be premature to say that the case is now solved for good.

Some of the suspects can be eliminated. Terrestrial vegetation, for one. Increasing terrestrial biomass in a glacial phase is unlikely, given that the advancing ice reduces the space on which plants can grow. The same holds true for most soils—though perhaps not permafrost soils.[8] The latter are exemplified by, say, the Siberian permafrost soils (locally called *yedoma*) which are notably carbon-rich (because organic matter in cold conditions is slow to decay). As these permafrost soils expanded their geographic range during glacial phases, it has been suggested that they might have locked in sufficient carbon to bring down atmospheric carbon dioxide levels—to release it subsequently back into the air as the permafrost shrank back.

The chief candidate to hide that carbon away is also the largest one: the ocean. And here the possibilities are several. Firstly, that the glacial world might store more carbon in biological, organic form, within either living plankton, or dead plankton drifting down to the sea floor. That seems unlikely, given that the glacial ocean does not seem to have been notably more biologically productive than the interglacial one. The extra fertilization from above caused by the greater amount of dust falling into the ocean was offset, it seems, by fewer nutrients from below, as a more effective 'lid' of surface water was formed by the meltwater runoff.[9]

Another way for the oceans to store carbon is simply as dissolved carbon dioxide—allowing more CO_2 to build up in deep waters during a glacial phase, say by an accumulation of gases produced by respiration of sinking organic matter—and then releasing it at the beginning of the next interglacial phase. There is a way of checking this process. The stored carbon dioxide would be expected to 'age' as it was kept at depth—that is, it would become progressively depleted in its short-lived radioactive isotope, ^{14}C, that is formed above the Earth's surface by cosmic rays colliding with nitrogen atoms in the atmosphere. In a deep, isolated store of carbon, the original ^{14}C taken into storage would diminish by radioactive decay, and, with no replenishment, the sequestered carbon would begin to age. Then, once released back into surface waters and the atmosphere, this 'old' carbon would be absorbed into, say, plankton skeletons as they form to give these an 'old' signature—that is, they would be less radioactive than expected.

For years, the hunt for such a store was fruitless, as samples of ocean floor sediments did not reveal the radiocarbon pattern to be expected from such a deep ocean store. Did the deep carbon store not exist at all? Recently, a strong candidate for just such a store was located, in the inhospitable waters of the Southern Ocean, adjacent to Antarctica.[10] There, in sediments of just the right age (marking the transition from the last glacial phase into the current interglacial), the tell-tale carbon pattern was found. Firstly, the lower part of the sediment core that represented the tail-end of the glacial phase included skeletons of foraminiferans that lived on the deep ocean floor: the carbon in these skeletons possessed an old signature, suggesting that these organisms lived in contact with a mass of water suffused with long-stored carbon. Then, in the sedimentary layers just above, marking the warming, were fossils of planktonic foraminiferans in which the old signature had appeared, indicating that the carbon store was now being released into the surface waters and into the atmosphere.

It is a neat story, and it might mean that the carbon mystery—or at least a part of it—is on the way to being solved. In science generally, though, finding a plausible, evidence-based solution to one question almost inevitably opens the door to another puzzle. The case of the missing carbon is no exception. Now that we suspect we know where the carbon has been hiding, what processes drove its capture and subsequent release?

This is still an open question.[11] How to trap carbon around the Southern Ocean? Expanding sea ice around Antarctica might hinder the winds from stirring and mixing the ocean—or perhaps the ice would simply be a physical barrier to carbon dioxide escape. And then, to reverse this process, one might invoke strengthening of winds—or alternatively changes in ocean currents triggered by meltwater influxes—to enhance ocean mixing and thus free the trapped carbon dioxide. Deciphering the mechanism of the beautifully regular and coincident Ice Age glacial–interglacial alternations will be a work in progress for some time to come.

There are, though, further levels of intricacy in the finely tuned climate machine of Ice Age Earth. Not all the climate oscillations may be ascribed to Milankovitch's marvellously effective mechanism. There are more—*very many* more—that show considerable regularity, that profoundly affected the Earth system and that remain, to a large extent, a mystery. Welcome to the crowded world of sub-Milankovitch cyclicity.

The millennial pattern

Let us, for now, focus on just one of the major glacial phases—and this would be most sensibly the one that separates us from the last warm interglacial phase, 125,000 years ago. In England this interval is called the Devensian Stage (though in Ireland it is known as the Midlandian); in western Europe it is the Weichselian (though in the Alps it is the Würm); in North America it goes by the name of the Wisconsian and in Poland the Vistulian, while in Siberia its alias is the Zyryanka, and in Chile the Llanquihe. No matter: it's all the same thing, seen through the terminological prisms of different national stratigraphic committees.

In length it is about 115,000 years (because the present, warm Holocene Epoch is a little over 10,000 years long)—and while broadly speaking the climate was cold, in detail it was *spectacularly* changeable. But to get a good look at this interval, one needs to know where to go, where the historical pickings (at this scale) are rich. The ocean floor sediments, mostly, won't do, because the burrowing organisms that live there have thoroughly scrambled the finest detail of the climate record. For a pristine history we need to go to long-lived lake basins, on the stagnant floors of which the sedimentary layers are little disturbed by burrowing. Or, most strikingly, one can turn to a stick of ice, some three kilometres high, drilled out of the centre of the Greenland Ice Sheet. That will do nicely, because over Greenland, with its high rates of snowfall, the entire ice thickness represents just a little over 125,000 years—so we have the lot.

With that stick of core, and after performing *many* analyses of isotope composition, dust content, and the like, the pattern of climate may be seen. Bracketed by the climate 'highs' of the last interglacial (called the Eemian in Europe, the Ipswichian in Britain, the Sangamon in North America . . . and so on, and so on) and of the Holocene one can discern about four broad peaks and troughs, corresponding to the 20,000-year wobble and the 40,000-year tilt cycles. As a whole, the trend through the entire 115,000-year interval is towards greater cold, and it is only in its last few thousand years (between about 20,000 and 15,000 years ago) that ice spread widely from the mountains into lowland regions.

But by far the most prominent features are a couple of dozen sharp, clear spikes in the signal, which are about half the amplitude of those separating glacial phase from interglacial phase. Is this simply analytical noise, the result of random fluctuations in oxygen isotope ratios? No—because drilling a parallel core in another part of the ice sheet gives essentially the same pattern.

These represent major, closely spaced climate fluctuations between the Holocene interglacial and the previous one.[12] They are called Dansgaard–Oeschger cycles, in honour of the two pioneering climate scientists, Willi Dansgaard and Hans Oeschger, who first recognized these features. There are twenty-six in all between the Holocene and the Eemian. They are somewhat irregu-

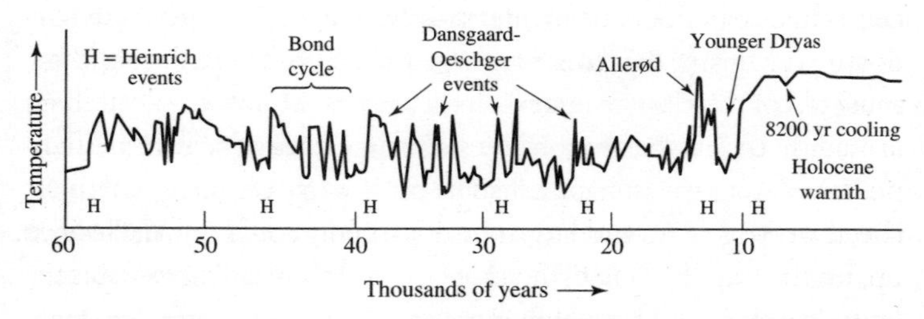

FIG. 24. Climate of the last 60,000 years. It shows the marked climate fluctuations of the second half of the last main glacial phase, punctuated by Heinrich events (partial collapses of the northern ice sheets), compared with the relative stability of the Holocene interglacial.

larly clustered, but where closely spaced they tend to recur every 1,500 years apart. The oxygen isotope data show that, between the cold parts of the oscillations (traditionally called stadials, to separate them from the larger-scale glacial events) and the warmings (interstadials), the average temperatures over Greenland changed by between 8 and 16 °, with abrupt warmings and rather slower coolings. This was a dynamic, rapidly changing environment, where climate instability was the norm.

The phenomenon is more byzantine still. These swings of warming and cold average a little under 1,500 years age—but seem not to be so very regularly spaced; this was not some finely tuned climate metronome, but a more skittish and irregular phenomenon. More, these cycles seem to form clusters, that have been called 'Bond cycles', several thousand years long, each showing a somewhat irregular cooling trend.[13] Something happens at the end of these cycles. To see what, we must move away from the ice and back to the ocean floor.

On the deep ocean floor of the north Atlantic, there are some places where the oozes have piled up quickly, so the burrowing organisms had less chance to chew through them. Evidence of these Dansgaard–Oeschger cycles can also be seen in these deposits—with a striking addition. Punctuating the fine oozes are distinct concentrations of sand, pebbles, and cobbles—the debris dropped from floating icebergs. The concentrations of debris are such that there must then have been enormous numbers of icebergs—'iceberg armadas', they have been called—drifting across the north Atlantic.[14] And, for that to happen, substantial parts of the northern ice sheets of the glacial world must have suddenly collapsed and broken up, for the fragments to drift across the ocean, reaching as far as the latitude of Spain today, before melting.

'Heinrich layers', these are called, after Hartmut Heinrich, the marine geologist who first discovered these curious deep-ocean strata. They are not a uniform phenomenon. The most typical and distinctive ones are characterized by limestone debris that has been traced back to the erosion of rocks in northern Canada, indicating collapse of the (then mighty) Laurentide ice sheet covering North America. With the ice came colder water, as shown by a curious but distinctive biological quirk. One of the common planktonic foraminiferans goes by the impressive name of *Neogloboquadrina pachyderma* (that is, the 'thick-skinned new *Globoquadrina*'). Its chambers are arranged in either a right-handed or left-handed coil, with the former being typical of warm water and the latter of cold water. They used to be thought of as just one species that had a temperature-controlled coiling switch, while now they are considered as two separate subspecies, occupying water masses of different temperature.[15] No matter: it is a fine temperature gauge, and the Heinrich layers show a sudden influx of the left-handed variety: along with the icebergs, the water had turned polar.

This story, though, is a tale of two hemispheres. The millennial-scale flips in climate show a world with intricate interconnections—and crucial regional differences. Look at the equivalent ice core of Antarctica, and you need to peer closely to see the close-set Dansgaard–Oeschger (D/O) pattern. It *is* there—but is far more muted, with temperature swings of only some 1–3 °C—nothing like the extravagant isotopic and temperature swings of the far north. But did they happen simultaneously in both hemispheres of the Earth, or was their timing offset?

One can match up the Greenland and Antarctica cores precisely, by using variations in the atmospheric methane content preserved within the ice cores. Fluctuations in this well-mixed but short-lived gas in the atmosphere act as precise time-lines between these two ice-masses, at opposite ends of the Earth. Using this chronometer, the warming and cooling trends seem to be systematically offset.[16] The world, at this scale, behaves rather like a climate see-saw, with one hemisphere warming while the other one cools down. More precisely still, they are seen to be *partially* offset (so not a fully symmetrical see-saw), so that northern cold coincided with southern warming, though intervals of peak warmth more or less coincided north and south. Carbon dioxide levels also oscillate in sympathy with these cycles—not 100 ppmv, as with the major glacial–interglacial cycles, but by about 20 ppmv—and in this case they most closely line up with the subtle Antarctic signal, rising as that continent warms and falling as it cools.[17]

How does it all work? There is a mechanism here, and (as usual) an underlying mystery. The basic pieces of this giant four-dimensional moving clockwork puzzle were put together in the 1980s by the great oceanographer, Wallace Broecker, and they are holding together pretty well.[18]

It is basically a matter of heat distribution—and of realignment of the Earth's radiators, the oceans, as these have much greater capacity to store heat than does the atmosphere. Currently, the Atlantic-facing countries of north-west Europe are several degrees Celsius warmer than they should be at their latitude, because they receive enormous amounts of heat carried northwards by the Gulf Stream. The Gulf Stream is just part of a longer, more intricate current loop or 'oceanic conveyor belt', and its motor might be said to be in the north rather than the south. In the far north of the Atlantic, the Gulf-derived waters, having given up their heat, cool, and thus become more dense. Furthermore, as the ocean surface freezes to form sea ice in winter, the adjacent water becomes yet saltier and denser, since the sea ice itself excludes salt as it forms. At a critical density threshold the cold, dense, salty waters sink, and begin a long journey southwards along the sea floor as what is called North Atlantic Deep Water, eventually reaching the Antarctic region. It is one of the returning loops of current that is the Gulf Stream.

The Gulf Stream is a marvellous machine to bring pleasant warmth to the far north, but it can be disrupted. Throw enough melting ice and melt water into the north Atlantic, for instance, and the salinity—and hence density—of the surface waters will

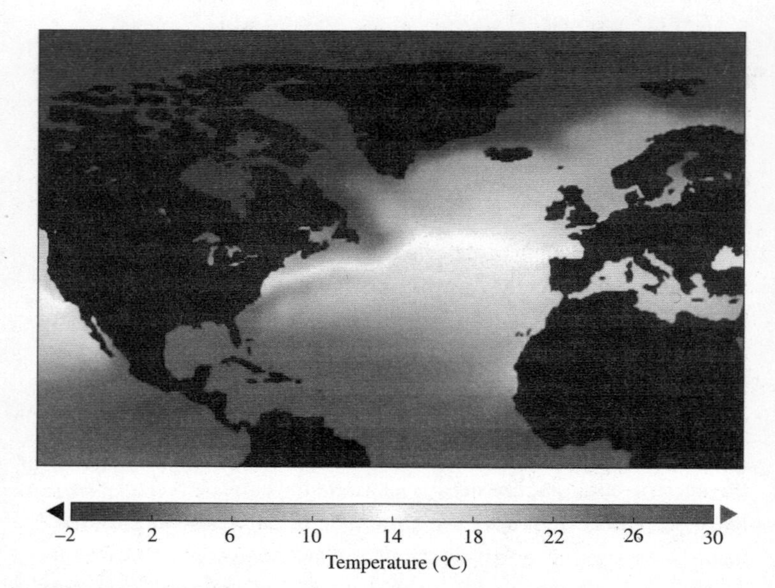

FIG. 25. The Gulf Stream is one of the strongest surface ocean currents and it delivers warm surface water from the Gulf of Mexico to the north-east Atlantic. It exerts a significant influence on the climate of Western Europe, with UK winters about 5 °C warmer than they would be otherwise.

be lowered sufficiently to prevent sinking in the far north. In stopping the formation of the North Atlantic Deep Water, the Gulf Stream is shut off, and north-west Europe can be thrown into a deep freeze. Heat then begins to build up in the deep south: Antarctica slowly warms, and—it seems—the Southern Ocean around Antarctica becomes a little less able to absorb carbon dioxide, and so concentrations of that gas in the atmosphere become a little higher.

This is a neat model, that in general still seems to hold up—but it does beg the question of what caused the meltwater/ice influx in the first place, and of the precise nature of teleconnections by which north and south are linked, not to mention the knock-on effects for other major climate systems elsewhere in the world, such as the tropical monsoon system.

The D/O and Heinrich events are obviously related, even though there are about three times as many of the former as of the latter. Nevertheless, the Heinrich events tend to occur at the end of the multi-millennial cooling trends (the 'Bond cycles'), and the aftermath of a Heinrich event is a particularly warm D/O cycle. One idea is that as the great northern ice sheets become larger and thicker, they would become a more effective blanket for the geothermal heat that escapes from the interior of the Earth; as this happens a greater part of them will begin to melt at their base, even as their upper parts remain frozen. As they begin to melt more and more at their base, the ice sheets lose their tight grip on the underlying rock surface, and can begin to slide catastrophically into the sea,[19] losing enough of their mass to cause global sea levels to rise by a metre or more—causing a Heinrich event and triggering a shut-down of the Gulf Stream.

There is some evidence for this, in that Heinrich events seem to have started once the northern ice sheets had grown to around their maximum size, a little over half a million years ago, as the 100,000-year world got into full swing. Other ideas, though, have been put forward for these kinds of pervasive millennial-scale climate changes, such as some kind of thousand-year rhythm in solar energy.[20]

The world as affected by ice

In thinking of the Ice Age, it is easy to take a mid-latitude view of things, especially if one hails from Europe or North America, and where those dioramas of mammoths and fur-clad humans and moving walls of ice have bitten deep into the subconscious. But, the shadow of ice reached further, as far as the tropics, as the changes swept across the entire world.

There is, though, some truth in the mid-latitude focus. For the polar regions have—thus far—largely stayed ice-bound through both glacial and interglacial intervals, at least wherever the polar chill combines with sufficient precipitation to form ice sheets. Northern Siberia, for instance, is cold enough for ice sheets to develop—but is too dry for ice to grow. There are temperature changes at the poles during the glacial–interglacial and stadial–interstadial transitions, as seen in those ice core records. In absolute terms the temperature changes here are greater than anywhere else on the globe—but nevertheless these are from icy to *bitterly* icy and back again.

The tropics have experienced the least temperature change. The equatorial region, during glacial phases, saw cooling of maybe 1 or 2 °C—but the tropics stayed essentially tropical as regards temperature. The debate here has focused mainly on other aspects of climate, such as humidity. Glacial phases, for instance, were generally drier across the Earth, as shown by the great increases in the transported dust arriving on the ice sheets, both north and south, during glacial and stadial phases. But how was this increased aridity expressed?

The central plains of China and of southern Europe, a little beyond the reach of the ice sheets, saw a succession of windstorms that draped the topography with thick layers of silt, known as loess.[21] As climate ameliorated in the interglacial phases, the silt supply slackened, and increased rainfall turned the silt surface into a soil—which in turn was buried under more silt as the climate turned colder once more.

One debate has been over what happened to the great tropical rainforests of Africa, South America, and Asia. Were they then as extensive, more or less, as they are today (or rather, were, before the current human-driven deforestation)? This might be suggested by their amazing biological diversity (again, before our recent depredations)—for how could this survive if their extent was massively reduced? Or did they shrink back as the rains failed, to be widely replaced by savannah, the biodiversity in this scenario surviving in wetter refugia on areas of higher ground?

This kind of question has been tackled by looking for places, such as long-lived lakes, where sediment accumulated over several glacial–interglacial phases and where the sediment layers contain pollen released by local vegetation. Such archives have not been easy to find. Even when found, they not easy to interpret, for the forest trees are abundant and diverse, but do not always release much pollen to wind, being more commonly insect-pollinated. Nevertheless, a picture of significant forest dieback has emerged.[22] This highlights the effectiveness of refugia—persistent forest stands—as preservers of biodiversity.

A yet more detailed archive is in limestone caves, where speleothems (stalagmites and stalactites) can grow over hundreds of thousands of years. They grow deep underground, in continuous darkness and—despite being seemingly isolated from the surface world above—can lock within them a history of changes in the climate above. Different patterns of climate can segregate heavy water molecules (those containing the heavy oxygen isotope ^{18}O) from light ones (those containing ^{16}O) to different degrees. In the monsoonal countries of Asia, a stronger monsoon system will produce rainfall with a greater proportion of 'light' water molecules.

While most of the rainfall washes off the ground to swell the rivers as they flow to the sea, a small proportion of it filters through the ground—some to drip down into underground caves where, infinitesimally slowly, the speleothems are forming. As these structures grow, year by year and century by century, the changing oxygen isotope composition within them reflects the strength of the monsoons as they pass above. What is more, the limestone of these growths can be directly dated because the speleothems, as they grow, incorporate tiny but measurable amounts of uranium into their structures, and the uranium, once in the rock, slowly but steadily decays into thorium,[23] the amounts of which can also be measured. Thus, the speleothems also contain within them a radioactive clock, charting their growth in years.

It needs a stable, well-ordered cave system to encourage the growth of a continuous, consistent speleothem climate archive. Caves in China, though, have now yielded complete records going back more than 200,000 years—well past the last interglacial into the preceding glacial phase. The best ones rival the Greenland ice sheet in the fidelity of their preserved history—and are also astonishingly similar in the details of their pattern.[24] The dominant signal is of major intensifications and decreases of the monsoon spaced about 20,000 years apart—the Earth's precession (wobble) signal, showing that (as predicted by calculation) when more summer sunlight warmed the continental interior, including the Tibetan plateau, there would be a larger updraught from the continent, more moisture-laden air drawn in from the sea—and therefore stronger monsoonal rainfall.

The cave records also show finer, millennial-scale rhythms—that appear to match closely, seemingly identically—with the stadial–interstadial cycles of the Greenland ice sheet. Here, the faster-paced monsoonal intensifications are linked with the interstadial warmings, while declines in the monsoons correspond to the stadials, the times when Atlantic circulation shut down—with the greatest declines linked with the 'iceberg armada', Heinrich events. Curiously, millennial-scale events are also present in the cave archives that precede the last interglacial, over 130,000 years ago—but these have a periodicity that is different from the better-known ones of the last glacial phase, at nearer 3,000 rather than 1,500 years. The difference is a hint that these mysterious 'sub-Milankovitch' climate cycles are more likely to be due to some kind of Earth-based control. Perhaps, therefore, they are linked with the volume of global ice (which would be different between different glacial intervals), rather than having some external cause (such as some kind of subtle 'pulsing' of the Sun's energy output).

Very many people depend on a regular monsoon to grow their food—but are threatened by one that is too strong. As we write, Pakistan is being devastated by abnormally strong monsoonal flooding of the 'Great Mother', the River Indus, that brings both good and ill, while the rains are wreaking havoc in China. It is clearly important to understand the controls on this phenomenon.

Where the winds blow

The world's climate patterns changed elsewhere, too, as the glacials and interglacials and stadials and interstadials of the Quaternary succeeded each other in the north. North-east Brazil today, for instance, is on the edge of something called the Inter-Tropical Convergence Zone, or ITCZ. It is where the winds coming from the northern and southern hemispheres meet, and it is essentially a rainbelt—visible on satellite images as a band of clouds. The climate in this part of Brazil is semi-arid: the rains come if the ITCZ comes far enough south, and do not if it stays to the north. Human history has seen a mixture of good years and bad years vivid enough to burn deep into the public consciousness, particularly in January, the start of either the rainy season, or of the feared *seco*—the drought.[25] On the evening of 13 January, for instance, the feast of St Luzia, one should put six salt crystals on a banana leaf and leave them overnight: the number of crystals that, by the morning, have dissolved in the dew are a premonition, it is said, of the number of months that will receive rain. One needs, after all, as much forewarning as one can get. Between 1877 and 1879 the rains did not come, and up to half a million people died.

The deeper history of these Brazilian droughts, like those of the Asian monsoons, can be read in speleothems. Caves in Brazil yield histories that, like the Chinese speleothems, are up to 200,000 years long, and also systematically datable by means of the uranium-thorium technique.[26] These give evidence that the climate here has been fickle on a much longer scale than humans have experienced. During prolonged, millennial-scale droughts the ITCZ stayed far to the north, and the stalactites and stalagmites ceased to grow completely. Their growth resumed each time the ITCZ came back south, and for several centuries the rains returned. The dating shows that the drought in this part of the world was broken as Greenland cooled and as the iceberg armadas of the Heinrich events spread across the north Atlantic: this was, thus, at the very time when the monsoonal rains weakened over Asia.

The effect of ice extended even further into the world's arid zones. The Sahara desert, for instance, is known to have bloomed in prehistoric times, for abundant, beautifully fashioned flint arrowheads can be found littering landscapes that are currently dry and barren. Earlier, these wetter intervals were termed 'pluvials', separating the drier or interpluvial episodes, and it used to be thought that they related directly to glacial and interglacial phases, so that as ice pushed down over Europe, the Sahara greened. The climate links, we now know, are more complicated than that. The ice core record shows that as the ice sheets grew large, the world became dustier and more arid, and the Sahara expanded well beyond its present extent, with stabilized desert sands lying beneath today's vegetation to the south of the present limit of active sand dunes;[27] so, more generally, the pluvial periods correspond to interglacial phases. The last greening of the Sahara took place as the ice sheets were shrinking back in early post-glacial times, and arid conditions returned once they had gone.

The world of ice in the Pleistocene was thus intricate and highly dynamic. It is time to consider an interval of climate that was considerably more stable: stable enough, in fact, to incubate an entire civilization. After the Pleistocene came the latest interglacial phase, that we call the Holocene.

Birth and Death of the Holocene

The human touch

It is just the latest of many climate phases of the Quaternary Period. The 103rd major shift in climate-driven global oxygen isotope values, to be precise, since the official-designated beginning of the Quaternary Period, 2.58 million years ago. And, many of those major phases, as we have seen, include dozens of climate oscillations far greater in scale than humans have witnessed since written records began. Nevertheless, it is *our* warm phase, that within which our civilization has grown, and hence it has been separated as a distinct epoch, the Holocene, a little over 0.01 of a million years long. Its counterpart is the Pleistocene Epoch, in which reside those other 2.57 million years of Quaternary time, and those other 102 major climate oscillations. Thus, we live—at least as far as formal geological nomenclature goes—in a privileged time.

When this epoch began, *Homo sapiens* had already existed for some 150,000 years. As a species its prospects might not have seemed bright: this creature lacked anything terribly impressive in the way of claws or teeth or thick fur or armour. But by being ingenious at developing what one might describe as artificial claws and teeth—axes and spears and arrows—it could kill and eat mammals considerably larger than itself. In those early days, it might not have prospered, exactly, but it clung to existence, seemingly weathering at least one very bad patch,[1] several tens of thousands of years ago, when its numbers dropped almost to extinction levels. It survived the climate oscillations of the late Pleistocene—the droughts and floods and episodes of bitter cold and killing heat—by adapting its behaviour or migrating as best it could.

Its migrations from its place of origin, Africa, were on an epic scale. The many thousands of individual and collective stories of hope, fear, endurance, courage, tragedy, and (less commonly) triumph are all lost. What remains is the evidence that humans, by the beginning of the Holocene, had spread widely over Europe and Asia, ousting (it seems) their kindred hominin species, *Homo neanderthalensis* and *Homo erectus*. As the ice retreated, they spread from eastern Siberia into North America, across the dry land (before the waters rose) of the Bering Strait. In frail boats, they had crossed thousands of miles of open ocean to reach many of the Pacific islands, and Australia too.

As the Holocene climate and sea level settled into something like comfortable stability, following the ups and downs of the late Pleistocene, some human populations found that a more certain and prosperous life could be had by settling to build communities and grow crops. And so another chapter in Earth history began. It is a chapter that will loom large. But first, the Holocene must be put into its place, as an episode of warmth not quite like any other.

A diversity of interglacials

There is a distinctive asymmetry about the pattern of successive glacial and interglacial phases. The 'glacial' intervals, as we have seen, are internally exceedingly complex. Depending on where in the world one is looking from, they showed close-set and more or less dramatic swings of temperature and rainfall. And they are *long*, lasting of the order of a hundred thousand years or more.

The interglacials separating these are different in character. Before we go further, we have to consider the business of words and definitions—and of their history. Amid the many climatic ups and downs of the ice ages, we define the interglacials as those episodes of peak warmth that most resemble the conditions of today. These interglacials, as those fabulous archives of ocean mud and Antarctic ice tell us, have been spaced at intervals of about a hundred thousand years, for the last million years or so of the Quaternary Period.

Interglacials were relatively brief by comparison with the glacial episodes. Quite how brief they were was a little hard to work out, on the basis of classical geological evidence. If one, for instance, finds a fossil lake deposit characterizing such or such an interglacial, and if it is just the right *kind* of lake, with seasonally changing rivers flowing into it, and with a stagnant, undisturbed floor, then distinctive annual layers of sediment may be preserved, and these may be counted. But it is exceedingly rare to find a complete lake-fill that silted up in such a well-behaved way, continuously through an entire interglacial phase. Nevertheless, estimates emerged that suggest that the last few interglacial phases—the ones for which the evidence was best preserved—had lasted of the order of ten thousand years or so—and so about a tenth of an entire glacial–interglacial cycle, or (to put it into the Milankovitch framework) about half a precession cycle.

The best estimates, naturally, came from our current interglacial phase, the Holocene. Here, one can take cores through the floors of lakes that are still silting up—and we also have the chronometer of radiocarbon that we have already alluded to in Chapter 7. This is a very useful timepiece for any materials that are younger than about 60,000 years, because carbon is present in a wide variety of materials—and particularly, of course, in organic remains. However, its action is not perfectly balanced, because the production of radiocarbon, ^{14}C, by the action of cosmic rays striking nitrogen atoms in the atmosphere, is not quite regular. This in turn is because the intensity of cosmic rays has been a little uneven through time, and also because some of the carbon that is absorbed by animals and plants may be 'old' and so lack the radiocarbon component. But, by applying sundry correction factors, useful dates can be obtained. And by these and other means, the Holocene was seen to begin a little more than ten thousand years ago. So, this fitted in with the inferred longevities of interglacial phases generally. It also suggested that we were coming to the end of the current interglacial, and that the Holocene, all things being equal, should soon give way to the next phase of glacial climate.

These estimates were not, as it turned out, too far out. A precise timescale of interglacial durations came with the first long core drilled through Antarctic ice, at the Vostok station. It penetrated a continuous 400,000-year record of compressed snowfall, thus capturing all the present interglacial and the three previous ones, and just the beginning of a yet older interglacial. Here, the comparative anatomy of the interglacials can be seen on one finely calibrated diagram, with time errors down to a matter of centuries.

In this Antarctic perspective, the Holocene turned out to be significantly the longest of these four interglacials. The previous interglacial, and the third one from the present, each had a brief warm 'spike' of about 4,000 years, and each was followed by a spell of somewhat lower temperature lasting some 10,000 years; in the first of these, at least, global temperatures peaked a little higher than today's, a little more of the world's ice sheets melted and sea level rose to perhaps some 5 metres higher than today's, based on evidence such as stranded shorelines on ocean islands.[2] The second interglacial from the present, by contrast, was a simpler temperature rise and fall lasting about 7,000 years. The Holocene—in the Vostok core—is distinct from these. It here appears as a stable temperature plateau, some 11,000 years long, emerging rapidly—but not quite abruptly—out of the depths of the last glaciation. Compared with the others, it is thus quite unusual, and represents the longest interval of stable warm climate, by some way, in the past 400,000 years.

This was the story for several years—until the next deeper ice core, through more than 800,000 years' worth of Antarctic ice (see Ch. 8) was drilled, analysed, and published. This borehole penetrated all of the fourth interglacial before the present one, of which only the tail-end had originally been captured by the Vostok core. This interglacial emerged as almost 30,000 years long (that is, the equivalent of about one and a half precession cycles) and thus much longer than the Holocene (so far). Also, like the last interglacial, it became a little (perhaps a degree Celsius) warmer, with sea levels peaking at perhaps 10 metres higher. It also resembles the Holocene in astronomical terms, and so it was suggested (tentatively, and controversially) that the Holocene *might* naturally, and without additional influences, last another 20,000 years.

The earlier interglacials, those more than half a million years old, each had its own character in detail too. But, compared to the more recent ones, they tended to be not quite as warm, yet that feebler warmth lasted a little longer, on average. Thus, on the whole, as the '100,000-year world' has evolved, the climate has become somewhat more extreme, somewhat harsher—and in itself that world is more extreme than the 40,000-year world that preceded it (Ch. 8). That *might* all be leading us somewhere, quite naturally—as we will discuss in the next chapter.

However, as ever in geology, we can turn back to history, and examine the beginning of the Holocene. It was far from being a simple onset of warmth.

Preludes to warmth

The history of the Holocene is distinctive among the recent interglacials in the way it started. This beginning cannot be clearly seen in the Antarctica ice cores—one can see a kink in the oxygen isotope

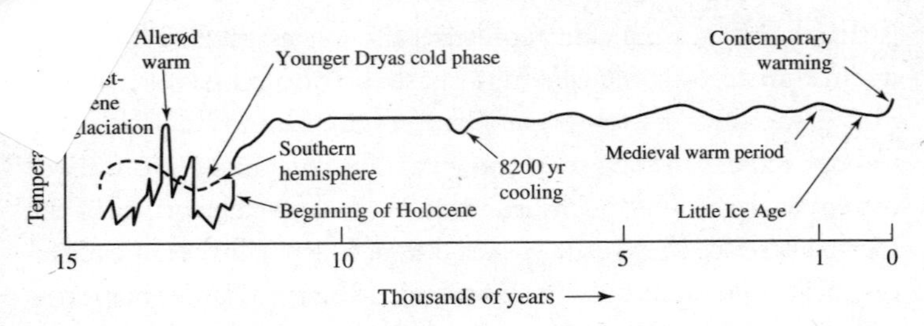

FIG. 26. Climate of the last 15,000 years, showing the complex transition (different in the northern and southern hemispheres) from the late Pleistocene glaciation to the Holocene interglacial, with its—so far—more muted climate changes.

curve there, as temperatures rise from the preceding glacial maximum—but it is subtle. To see the transition at its clearest, one needs to turn to the north, and in particular to the Greenland ice sheet, where the history of events was different.

There, climate history rode a switchback.[3] Coming from the Glacial maximum, temperatures suddenly rose, 14,700 years ago, with the average temperature of the north Atlantic and surrounding areas increasing rapidly by some 5 °C (over Greenland, the temperature hike approached 20 °C). Temperatures remained around these levels for nearly two millennia—and then fell suddenly by a similar amount, as the whole region went into a deep freeze once more; in north-west Scotland, for instance, which had thawed completely in the warm episode, a new ice sheet began to grow, so that this cold snap used to be called there the Loch Lomond Re-advance. The cold interval is now widely referred to, though, as the 'Younger Dryas', named after *Dryas octopetala*, an alpine plant that bloomed across Scandinavia as the chill set in.

The arctic temperatures persisted for another thousand years, and then, 11,700 years ago, temperatures suddenly soared again in another climate transformation—only this time the warm temperatures stayed, and this transition marks the beginning of the Holocene.

That beginning is now defined, formally, in a Greenland ice-core,[4] and the ice layers there show the abruptness of this climate revolution—at least in the northern hemisphere.

How sudden is sudden? These climate boundaries have been examined exceedingly minutely in the Greenland ice-cores, and the oxygen isotopes suggest that the temperature rises were accomplished in a few decades—perhaps in one human lifetime. That seems quite fast enough. But the reorganization of the climate system (which determines the source of the moisture to Greenland) can be tracked by looking at the proportion of heavy hydrogen (deuterium) to normal hydrogen in the ice. And here, the change seems to have been accomplished in something between *one and three years.*[5]

It appears akin to those other millennial-scale climate jumps so commonplace in the last glacial phase, the Dansgaard–Oeschger cycles. To bring about the initial warming, the water south of Greenland must have been dense enough to cool and sink to the ocean floor, to start the climate loop that would eventually come back as the warmth-giving Gulf Stream. Then, something must have switched off the Gulf Stream, before the current once more reasserted itself.

But what, here, was the switch-off mechanism? As this is the youngest of these events, it is also the one for which the evidence is best preserved. As with the earlier D/O events, a plausible mechanism is to flood the Atlantic with glacial melt water to create a low density 'lid' to prevent the sinking of dense masses of water. How, now, does one flood an ocean?

Human timescales are short—*very* short on a geological timescale, and therefore most of us witness only a range of common geological phenomena, and not the rare, extreme examples. With volcanoes, for instance, we might remember—some of us—the awesome images of an eruption such as, say, Mount St Helens, which, throwing a cubic kilometre of ash suddenly into the air, devastated the local landscape and created all kinds of mayhem. We might also have read the accounts of, say, the cataclysmic eruption of Krakatoa, some three generations earlier, which was a larger and much more lethal event, producing about 50 km^3 of ash. But, no modern humans witnessed the massive Yellowstone eruptions of about a million years ago: each of these poured out of the order of 1,000 cubic kilometres of ash, which fell over most of the USA.

Similarly, with floods, we have the television and satellite images of the Pakistan floods before our eyes, taking homes and livelihoods away from millions of people (and lives away from thousands). But some of the floods of the past were off any scale of modern-day comparison—and so are difficult to entertain, even when the evidence of their action shouts out at you.

Just after the First World War, the geologist J Harlen Bretz (there really is no full stop after the J, and if one was inadvertently added in his presence, then by all accounts his wrath was a terrible thing to behold) was surveying the eastern part of Washington State, in the USA. The landscape for thousands of square kilometres around was bizarre, extraordinary. It was criss-crossed by canyons hundreds of feet deep, which cut through soft surface silt and hard basalt bedrock alike, and that were interconnected in such a way that water must have flowed over the rims of the canyons and eroded spillways that interlinked them. The channelled scablands, it's called. There were enormous potholes, too, and ripples—like the sand ripples one sees on a beach, but these were tens of feet high and made, not of sand, but of gravel and boulders: they were so big, indeed, that he initially did not recognize them as ripples, even when he was walking across them (it took a view from an aeroplane to recognize them for what they were). As he realized what was going on around him he must have felt like the Incredible Shrinking Man, set to do battle with spiders and caterpillars. For he drew what seemed to him the only sensible conclusion, that he was walking amid the aftermath of a flood—a flood of monstrous, Biblical proportions.[6]

The geological community of the day, infused with notions of gradual change, fought this concept tooth and nail. To be fair, Bretz did not always help his own case, initially arguing for a single event—it turned out that there had been dozens—and being vague about the origin of the flood. Nevertheless, the evidence upon which he based his case was clear. Yet it took some thirty years for the idea to be generally accepted (and Harlen Bretz, happily, lived to see his idea triumphantly vindicated). Megafloods to dwarf anything seen today have been a feature of the geological past, in the channelled scablands and elsewhere (enormous examples have been found in Siberia, for instance). They are linked to the melting of continental ice sheets, typically by the formation of a very large lake held in place by an ice barrier, that then drains catastrophically when the ice dam fails. The ensuing flood can change the world.

A meltwater flood of this sort has, for years, been seen as the most likely trigger for the abrupt reversal of climate of the Younger Dryas, suddenly placing a low-density, current-stopping lid across the north Atlantic.[7] But where did it come from? As the Laurentide ice sheet covering north America was melting, Lake Agassiz—larger than all the current Great Lakes together—lay next to it. The fertile lake silts and beach deposits that mark its successive margins are scattered across much of Manitoba, western Ontario, northern Minnesota, eastern North Dakota, and Saskatchewan—over 440,000 square kilometres in total. Containing, at its maximum, probably more water than all the world's lakes today, it is a temptingly large source of fresh water with which to stop the circulation of an ocean.

But just how did the water reach the sea? There are a number of possible escape routes that could suddenly open as the ice barriers fell: southwards, ultimately through the Mississippi; eastwards through the St Lawrence Seaway via Lake Superior; or west and north, into the Arctic via the Mackenzie River. For years, the eastern route, via the St Lawrence, was seen as the most likely. But, frustratingly, there was no physical evidence of the right age there—no scoured channels, no boulder fields.

Then, in the far north of Canada, by the Mackenzie River, evidence was found of widespread scouring and spreads of gravel and boulders, which could be dated precisely enough to fall into the 12,900-year bracket—and thus as near to the beginning of the Younger Dryas as could be hoped for.[8] This injection of several thousand cubic kilometres of water straight into the Arctic Ocean is now the prime suspect for triggering a millennium of northern hemisphere cold.

It is a dramatic enough way to alter climate—but not the most dramatic that has been proposed. A comet-burst over the northern hemisphere might—it has recently been suggested—have caused the climate change at the beginning of the Younger Dryas, and for good measure also an extinction event of large mammals (including the mammoths) and the end of the distinctive 'Clovis' archaeological culture of North America. It is not simply a fanciful idea: evidence cited includes the widespread presence in North America of a 'black layer' equivalent in age to the beginning of the Younger Dryas and rich in charcoal, soot, iridium-rich grains, and nanodiamonds. The cataclysm of explosions and wildfires is seen as having disrupted both ecosystems and human society, while the effect on climate was described as like a nuclear winter, with an injection of dust, sulphates, nitrogen compounds, and water vapour into the atmosphere, blocking out sunlight and then being prolonged by feedback effects.[9] It is a captivating idea and was greeted with interest—but also with much caution. Extraordinary hypotheses demand extraordinary evidence, after all. Time, and more evidence, will tell. Our money (for what it is worth) is still on the megafloods.

Shaping the Holocene world

Climate defines the shape of the world. It does this quite fundamentally, on a large—indeed literally on a global—scale. We are dealing with much more, here, than the distribution of the belts of cold and hot and temperate climate, or of the patterns of rainfall that determine the extent of jungle and semi-desert, savannah and boreal forest. We are dealing here with the production of landscape and—in this particular instance—with the creation of a stable platform for humanity. This is a quite tangible platform, for countless human feet to walk across in the search for a survivable present and a prosperous future. This platform has been shaped and is still controlled, quite literally, by the way climate governs the level of the sea.

Twenty thousand years ago, at the height of the last Ice Age, the Earth's land masses were larger. The ice sheets had grown to a maximum: those of Antarctica and Greenland, others covering northern and central Europe and North America, and yet others capping the world's mountain belts. Enough water had been extracted to bring the sea level down to some 130 metres below its present level. Humans walked then across the North Sea and across what are now the flooded continental shelves of the world.

It was not, though, an immobile world, simply to be concealed or revealed as the level of the ocean waters rose and fell. The weather machine was in continuous action, eroding the exposed land surface, the resulting sedimentary detritus being carried down to what was then the new shoreline, far below the present one. And among the softest and most easily erodible sediment was that which had formed around still earlier, higher shorelines as beaches and deltas. These short-lived structures were dismantled by wind and rain and river-flow into their constituent particles and swept down into the shrunken oceans.

As the world then warmed, and the ice began to melt and pour back into the oceans, this machine for engineering a global landscape was fundamentally restructured. The transition to the Holocene interglacial phase had been swift—dramatically so in the northern hemisphere. The reaction of sea level was slower, though, for it takes time to melt what would ultimately be some 10,000 cubic kilometres of ice.

The sea level rose continuously for the first five thousand years of the Holocene, as the world's ice sheets decayed back. Not quite steadily, though: there were intervals when that sea level rise was fast enough—over 45 mm a year—to drown coral reefs.[10] During those times, the sea surface (and hence life-giving light) moved upwards and beyond the reach even of the fastest-growing reef-building corals, such as the Elkhorn coral, *Acropora palmata*. One of those pulses, 8,200 years ago, marked the final draining event of Lake Agassiz, raising sea level by perhaps a metre at one go. It did not bring the post-glacial thaw to a complete halt to produce a second Younger Dryas event—but it has been implicated in slowing north Atlantic circulation sufficiently to produce a distinct, brief chill, of some 300 years, across the northern hemisphere.[11]

By 5,000 years ago, the sea level had reached its present level. There is still debate over whether the sea level has subsequently stayed essentially the same, within a metre or so, or whether there have been slightly greater oscillations (of up to, say, a couple of metres).[12] The uncertainty arises simply because of the difficulty, at this fine scale, of distinguishing changes in global sea level from changes in the land, particularly as the crust has readjusted to the addition and then removal of enormously heavy masses of ice (that in some parts of the world depressed the Earth's crust by half a kilometre or more). No matter: geologically, this is the very smallest of small changes as regards sea level. With the shoreline to all intents and purposes constant, the world's geography could adjust to it.

The world's rivers, yet to be tamed by humans, poured sediment down towards the sea—not least, at mid-latitudes, through the erosional reworking and transport of the enormous amount of debris left by the departing ice. Much of this sediment amassed at and around the shoreline as beaches, silting-up estuaries, deltas, and coastal plains. The scale and speed of this geomorphological construction is easy to forget, particularly given the brief timescale (and great distractions) of a human life. The landscape built in this way, in just the last five thousand years, forms most of Bangladesh and the Netherlands; a good deal of Belgium and the Fenland of eastern England; the whole of the Mississippi delta (comprising most of Louisiana); the Po delta, including Venice; much of Vietnam, from Haiphong up the Red River to Hanoi; the Nile delta including Alexandria; and in China, Shanghai. One can go on, almost endlessly, around the world's coastlines, picking out examples large and small.

It is a world that has physically adjusted to a very particular and remarkably stable climate state. And the new—and very extensive—additions to the terrestrial landscape, all at or within a very few metres of sea level, are the parts of the world that have, by and large, been most amenable for human life. On these, there is an abundance of water, deep nutritious soils, and flat, easily navigable landscapes. These have been perhaps the key feature in allowing the growth of a numerous human population and complex civilization, and they remain, acre for acre, the most densely inhabited and valuable parts of the world, utterly indispensible to maintaining present human numbers on the planet. Their position, though, as we shall see, is a little too precarious for comfort—as are those, therefore, of their human passengers.

Such concerns aside, what we have with such landscape creation is a wealth of evidence: a surfeit of strata rich in signals of climatic and environmental history, more abundant by far than those of previous interglacials and (by contrast with the ice core and deep ocean evidence) giving clues, around the world, of a variety of local climatic conditions. There is a quite extraordinary range of clues to past climate buried within those near-surface sedimentary layers, and some of those histories are now being measured in years rather than centuries and millennia.[13] There are the trunks and leaves and stems of plants—and their hard-wearing pollen grains too—and other plants, particularly the microscopic, environmentally sensitive silica diatoms. Among invertebrates, there are the ostracods (the 'seed-shrimps' we also met in ancient Antarctic Lake Boreas), smaller than a rice-grain and fabulously abundant in ponds and lakes, and snails and beetles (especially their tough wing-cases or elytra) and also the chironomid flies (midges) whose remains are preserved surprisingly commonly. Among vertebrates, it is the large, impressive beasts—deer and wild boar and wolf—that get the headlines, yet the smaller ones—the mice and voles and shrews—whose remains (especially their teeth) are usually more common and generally more useful. All these creatures have environmental and climatic tolerances and almost all are still alive today, so from their preserved remains one can help reconstruct the passage of climate in the Holocene.

There are stranger assemblages of evidence, also, as we will see. One particular type of evidence, though, that comes into its own in the Holocene, is the use of tree-rings, which has its own science (dendrochronology) and specialist practitioners. Tree-rings not only provide a measure of the passage of years, but also a record of whether those years were good or bad for growth, the latter by reason of cold, drought, or other misfortune. By cross-matching particular patterns of good years (wide rings) and bad years (narrow rings) as a kind of bar-code, one can extend a tree-ring history back from the present day, via older and older fossil trees, back to the beginning of the Holocene. This is not quite as simple as it sounds, for individual trees can have good and bad years from quite local luck and misfortune that have little to do with climate—such as ground disturbances or changes in nearby stream courses or the depredations of deer or beavers. The random, locally influenced patterns of wide and narrow have to be separated from those produced by climate to enable regional environmental histories to be put together. And tree-rings cannot provide data on those lands, far north and far south, where trees do not grow (while tropical trees, with their more constant year-round growth, have not provided very much useful information). So: they are a wonderful source of information but, as with any other form of climate evidence, one with significant limitations; one seeks to overcome these, at least in part, by meshing with other forms of evidence that possess different limitations.

Patterns in Holocene climate

The ice-core records show the Holocene to be a climate plateau longer and more uniform than any, warm or cold, in the past 400,000 years—and this is borne out, too, by the way sea level has stayed remarkably constant since its level stabilized, some 5,000 years ago.

Yet, while this overall picture has stayed constant, looking in detail (something that is much more possible than with previous interglacials) shows variation on different scales—variation significant enough to represent the difference between life and death for human communities, as regards temperature and rainfall, the prevalence of floods and storms, the likelihood of extreme events, and so on.

At the broadest scale, in some regions there is a clear difference between the early and late Holocene. There is, for instance, the greening of the Sahara during the early Holocene, already alluded to, and its subsequent drying about 4,000 years ago. There has also been a clear change in the behaviour of the Asian monsoons during the Holocene, as deduced from the isotopic signals they left in stalactites and stalagmites growing deep underground (see Ch. 8). There was greater rainfall early in the Holocene than now, sufficient for the Indus River of the early part of this epoch to be described as a 'monster river'.[14] The broad pattern here seems to follow an astronomical trend—the pattern of summer sunlight at the latitude of China that decreased steadily through the Holocene, plausibly to produce less strongly rising air from the continent—and therefore less moist air brought in from the sea.

Superimposed on this broad trend are finer-scale changes, with phases of weaker monsoon rains spaced at roughly thousand-year intervals. These times of 'failed monsoons' seem to link in time (and by inference, in process) with the episodes of chill recorded in the Greenland ice sheet that include the widespread '8,200-year event' probably driven by the final outburst from Lake Agassiz. So continues the association, from the preceding glacial phase, of cooling events and circulation slowdown in the Atlantic, with failures of the Asian monsoon.[15]

A later failure of the Asian rains, 4,300 years ago, seems to be a more local feature, without a clear north Atlantic connection. However, it took place as humans were beginning to create settled agricultural civilizations. Such a society was more obviously vulnerable to the vicissitudes of climate, and also created an archaeological record that can be compared with the geological record of climate. As that drought took place—and plausibly *because* it took place—the Neolithic culture in China collapsed, as did those, it seems, in Mesopotamia and India.[16]

There are other means of tracking Holocene climate—some quite unexpected. In the arid and semi-arid parts of the mid-west of the USA, for instance, there live families of packrats. Members of the family Muridae, they have large eyes and ears, strong feet and long tails, usually weigh in at a couple of hundred grams and have pale brown to almost black coats. In short, they are remarkably cute. They are also good collectors of palaeoclimate information because they constructed effective and durable climate recorders. They build nests, termed middens, in caves and rock hollows, by foraging for sticks and twigs and leaves that they bring back and cement into place using, rather indelicately (to human rather than packrat society, that is), their own urine. This is a rather viscous substance that dries to a glossy material called amberat.

The nests continue to be added to and embellished by this means—they can grow to be large, impressive structures. Cemented into them are the remains of packrat meals—leaves, seeds, insects, bones. These represent a selection of the surrounding flora and fauna, and the amberat is a fine preserving medium. Delicate structures can be preserved for tens of thousands of years, and the material thus fossilized can be carbon-dated, to show when particular climate-related assemblages of animals and plants lived.[17] On the west coast of the USA, for instance, the handiwork—or paw-work—of the packrats shows that the early Holocene was cooler and wetter than today, with hotter, drier conditions developing later.

The packrat has, in terms of palaeoclimatic fidelity, a cousin and rival in south-west Africa—the rock hyrax, which can also be considered cute, if you like your rodents to be generously proportioned. Hyraxes also make middens, but these are even more indelicate than those of the packrat, being essentially multi-generational family latrines, consisting of many thin layers of faecal matter cemented by urine. This makes them not quite so good as regards a high-fidelity fossil record (since the evidence has passed through the whole of a hyrax, as it were). However, they are much better as regards time resolution because, unlike the packrat middens, which are a bit of a geometric mess, the fine layering gives rise to detailed stratigraphies that, spliced together with the help of radiocarbon dating, can span thousands of years.[18]

The hyrax record, like that of the packrats, shows a succession of climatic states. These states are not inferred this time simply from palaeoecological analysis of the midden-bound fossils themselves, as much of the detail has been digested, but from the subtler chemical signals of vegetational change. The carbon atoms in the midden layers, for instance, show systematic changes in the proportion of light (^{12}C) and heavy (^{13}C) stable isotopes, and these variations reflect the changing vegetation that formed the hyrax food supply.

Plants today can be divided into two broad groups depending on the way they chemically process the carbon they have taken from the air in photosynthesis. There are C_3 plants, the earliest and still the most abundant form, accounting for some 97 per cent of plant species; their carbon metabolism in photosynthesis leads initially to a three-carbon molecule (hence the term) and tends to favour the take-up of light (^{12}C) rather than heavy (^{13}C) carbon atoms. And there are C_4 plants (mainly certain types of grasses)—a newer invention, that arose around early Oligocene times, more or less when ice started to grow on Antarctica. A glacial world is in general a drier world, and the C_4 molecular machinery is generally more efficient at coping with drier conditions. These plants incorporate a larger proportion of heavy carbon atoms in their tissues. The relatively few C_4 species thrive where conditions suit them, and now make up about 25 per cent of the world's primary plant productivity.

The hyraxes take these general patterns and adapt them (subvert them, one might say) to their local conditions and dietary preferences. Given the choice, these animals will feed mainly on grasses, including C_4 ones, but in times of drought, they make do with a tree- or shrub-based (C_3) diet: the latter plants, with their longer roots, can tap into deep groundwater more effectively than can the grasses.

The midden material may also be analysed to see differences in the proportions of light (^{14}N) versus heavy (^{15}N) nitrogen isotopes. The processes governing the shuffling of nitrogen atoms in the surface environment are complex; but, empirically, drier conditions have been shown to lead to higher proportions of the heavier nitrogen isotope in animal and vegetable material—for both C_3 and C_4 plants.

Thus the plant remains, processed year by year by the hyraxes and deposited in their well-organized latrines, show, via the patterns of their carbon and nitrogen isotopes, the passage of climate across south-west Africa. As with the Asian and western USA climate archives, the Holocene, very broadly, was marked by generally wetter conditions in its early part to generally drier conditions later. Within this lies a succession of drier and wetter events; major events are broadly millennial-scale, each lasting several centuries, while they have shorter, decadal-scale wet–dry cycles superimposed on them.

How did these events fit in to wider patterns across the world? Most obviously, the land became dry as the adjacent ocean off western Africa became colder, this marine chill being seen in an increased proportion of that distinctive cold-loving, left-coiling form of the planktonic foraminiferan *Neogloboquadrina pachyderma* (see Ch. 8) falling onto the sea floor. The sea became colder there because cold, deep waters were then upwelling more strongly near the African coastline. This upwelling was driven by strengthened trade winds, which in turn are influenced by shifting patterns of ocean currents in far-off regions of the world—as we saw also for variations in the Asian monsoon. The land and sea and air and ice together form an intricate machine. It has driven particular patterns of recent Earth climate, as shifts in the balance of complex global linkages between these domains have led to changes that could spell the difference between times of plenty and times of famine for emerging human societies.

There are other measures of climate besides temperature and humidity. Storminess, is one, for instance, though it is problematic to measure. Hurricanes may pass across an area to cause devastation and lay waste to forests—but on land the trees soon grow back, and little may be left as evidence. But, if there are periods of increased storminess, then more sediment may be washed into ponds and lakes by successive storms, and accumulate there as thick, distinctive strata. By measuring the numbers of such storm-generated sediment layers, and placing them into a time framework, the relative storminess can be inferred. It is necessary to look at a lot of lakes to obtain a robust pattern, but this exercise, carried out for the Holocene of north-east North America, suggests that the region saw peaks of storminess, spaced some two to four thousand years apart.[19] What caused them? The timing seems to be different from the variations in the Asian monsoon. In this case a link was suggested to long-term changes in patterns of high and low pressure over the Atlantic region (the North Atlantic Oscillation, it is called). Farther south, however, hurricanes making landfall on Texas seemed to show no obvious pattern,[20] the contrast being a good demonstration of the complexity of regional climate patterns.

The last thousand years

As we approach the present, the physical, geological evidence of climate state becomes yet more abundant, and is joined by written evidence of human scribes, noting good harvests and bad harvests, droughts and floods, mild or hard winters and balmy or dreadful summers. The picture becomes busier, rather like a series of pointillist paintings, each with ever more crowded dots and dashes of paint—though the overall pattern may often remain a little too impressionist for the comfort of scientists seeking precision and rigour. For instance, the written historical data, abundant as it is, was until very recently poetically descriptive rather than precise and numerical: it is hard to translate descriptions of, say, a particularly cold, sodden medieval harvest-time into the kind of data needed to construct a numerical climate model. For all the cultural advances of civilization, systematic temperature measurements and rain-gauge data only started to be taken in the last couple of centuries—and then, initially, only in a few parts of the world.

To put together a picture of the last millennium, we need, therefore, first to incorporate as many lines of evidence as possible—from pollen and insects in lake sediments, tree-rings, ice-cores, stalactites and stalagmites, written archives, and more. We also need first to concentrate on one part of the world, or at least one hemisphere (as we have seen that the two hemispheres behave differently, even at much coarser timescales). For the northern hemisphere—where most work has been done—each of the individual lines of evidence gives a slightly different pattern, and so together they define a kind of 'envelope' of inferred temperatures through time.

Within the envelope (or at least within its individual components) there are ups and downs on the scale of years, decades, and centuries. But the overall pattern is for slightly higher temperatures early on, in medieval times—the eleventh to fourteenth centuries—diminishing to colder temperatures that then dominated the seventeenth to nineteenth centuries. The difference in average annual temperatures between these two states is generally estimated at under one degree Celsius—and most studies suggest less than half a degree. Nevertheless, that difference in climate was palpable.[21] These were the Medieval Warm Period and the Little Ice Age, and this is where climate statistics take on a human dimension: the translation is more vivid than the simple numbers might indicate. The Little Ice Age, for instance, seems to have killed off the Viking colony in Greenland in the mid-fifteenth century, after the warmth of Medieval times had tempted the Vikings to settle there.

The depth of the Little Ice Age is typically depicted via charming scenes of cheerful inhabitants of London and Amsterdam skating on the frozen Thames and Amstel rivers. Charles Kingsley, though, writing in *Madame How and Lady Why* (subtitled as *First Lessons in Earth Lore for Children*) saw a different side. He recalled standing on the Battersea Bridge in the great winter of 1837–8, around which enormous slabs of ice were 'leaping and piling up on each other's backs' with the bridge itself shuddering and trembling from the shock of their impacts, to the sound of the grinding and roaring of the ice-masses. To feed the imaginations of Victorian children, he mused, then, how it must feel to be among clashing Arctic ice-floes, as an entrapped ship might be squeezed and lifted upwards. This was Siberia paying a visit to London, as a hemisphere's average temperature dropped by less than a degree.

The Sun's heartbeat?

There is another actor involved in this play of the world's climate. It is uncertain, though, whether the role is a starring one, or that of a minor player. In some ways this actor is the most obvious one of all. It is the Sun.

Variations in the Sun's output have been used to explain patterns in Earth climate history. Indeed, we know that, at the very longest timescale, the Sun is slowly getting brighter and hotter, and in this book we have had to seek to explain why the Earth was not continually frozen in its early days, when the Sun was young and faint (see Ch. 1).

It is much shorter timescales, though, on which the Sun has been most strongly implicated in climate change. The clearest expression of changes in the Sun's activity is in the rhythmically changing number of sunspots on its surface, in which an 11-year cycle is dominant. Sunspots are dark—and hence relatively cool—patches, which are thought to act as 'thermal plugs', literally keeping the heat beneath them confined inside the Sun. However, sunspots are also associated with brighter (and hotter) patches, called faculae, and overall the times when sunspot numbers are high coincide with a slightly greater output of heat and light overall from the Sun—but by only about one part in a thousand.[22]

Superimposed on the 11-year cycle are longer, century-scale, variations, when sunspots are generally more or less abundant. The most famous of these, and the first that was directly observed, was a 70-year interval, roughly between AD 1645 and 1715, when there were virtually no sunspots at all. It is called the Maunder Minimum, named after the nineteenth-century astronomer Edward Maunder, who discovered this episode of solar inactivity by combing through old observatory records.

The Maunder Minimum lies within the Little Ice Age—and has been suggested by some to be at least partly the cause of the cold temperatures of that time.[23] However, the change in solar radiation between sunspot highs and sunspot lows is, as we have noted, very slight.

So, another mechanism has been sought, that could feasibly have a perceptible effect on global temperatures. The most plausible of these is the effect of solar radiation ('the solar wind') on cosmic rays, which are *extremely*[24] high-velocity particles—mostly protons that speed through outer space, the aftermath, it is thought, of supernova explosions. The Earth is normally partly shielded from cosmic rays by its magnetic field, and this shield is strengthened when the Sun is more active—that is, when it has a greater number of sunspots.

Cosmic rays, though, passing through water vapour, can make that vapour condense into trails of tiny water droplets. This is the principle on which the old-fashioned cloud chamber in physics was constructed in order to reveal the tracks of moving ions. Thus, cosmic rays have been considered to be cloud makers. The more that get through to the atmosphere (in times of fewer sunspots, and *particularly* in times like the Maunder Minimum), the cloudier—and therefore cooler—the Earth becomes.

There are a lot of unknowns here, including quite how clouds are created by cosmic rays (there is currently an experiment taking place at the particle accelerator facility at CERN, handily called CLOUD, that is studying this). It is a topic of lively debate: few doubt that the Sun's behaviour can influence climate—the question is whether it has a *substantial* influence on climate.

The cosmic rays provide more evidence for the debate, in that they not only help (to some extent) to create clouds, but they also smash into atoms of oxygen and nitrogen to create different kinds of atoms—chief among these being the short-lived radioisotope carbon-14 (created from nitrogen, with a half-life of 5,730 years) and the longer-lived beryllium-10 (created from nitrogen or oxygen, with a half-life of 1.5 million years). Carbon-14 can get incorporated into the wood of growing trees. Comparing the amount of this isotope with the age of the tree (by counting the rings) can reveal times when either more or less carbon-14 was being produced, therefore acting as a kind of a historical record of solar activity. Beryllium-10, mainly when it falls on to the polar ice-caps and is trapped in layers of snow and ice, has been used in a similar way.

The results of these studies are not always consistent with each other, because the paths that carbon-14 and beryllium-10 take to their final resting places in wood and ice are not quite straightforward. However, the record of these isotopes generally mirrors the historical observations—and extends them into times before sunspots were observed. It appears that the Maunder Minimum and earlier such events broadly coincide with cold phases, and times of higher sunspot activity broadly seem to coincide with warmer phases. There are also indications (albeit ones that have been strongly disputed) that the last half-century or so has seen the highest sunspot activity of the last millennium. So, could the marked warming of the last half-century be simply due to increased solar activity?

This seems unlikely.[25] The only way that one can *really* accurately measure the total radiation from the Sun is to send up a specially designed satellite, away from the blurring influence of the atmosphere and of the clouds and dust within it. This has been done since 1978, and so there is now over thirty years' worth of data. This has shown three sunspot cycles, peaking around 1980, 1991, and 2002, with minuscule differences from peak to peak and trough to trough, meaning that on average the Sun has held steady. In that time, global temperatures have undergone a steep climb, rising by about half a degree Celsius.

Changes in the Sun, therefore, do not seem responsible—or not *largely* responsible—for recent temperature change. That is not to say that there isn't a great deal to learn about the workings of the star that sustains us. Could it, perhaps, have undergone episodes of 'hyperactivity' in the past—variations in activity that exceeded those we have seen in recent historical times? Such things have been mooted, but the evidence to date is scant. The physicists and the climate scientists between them will probably turn up some interesting stories in coming years.[26]

There are other fluctuations in climate, though, that are considerably less well-behaved than the sunspot cycles—and more disruptive. Take the weather-children of the Pacific, for instance: when in full cry, they certainly can't be ignored.

The child that never grows old

In Neverland, Peter Pan never grows up. In the workings of global weather, there is also a perpetual child, fearsome and capricious in equal measure. He is called 'El Niño' by the people who live on the west coast of the Americas, and he appears every few years—usually around Christmas—to turn their lives upside down with flood and famine. El Niño has a sister and opposite, La Niña, who rivals him for spite, as she is the bringer of drought. Between them, these two weather-children (and the normal years that intersperse their appearance) bring years of plenty, and lean years and, sometimes also disaster years. They have a wide reach. The whole of the Pacific—and Africa too—is their playground.

It would be prudent to know them better. But as they are very old, and human history is short, we need to look elsewhere for witnesses to their behaviour.

One such witness to the passage of good and bad fortune among fishermen and farmers, and all who depend on them, lives a slow and long life—up to a few centuries, with luck—amid the sunshine and waves of a Pacific island. As it grows, it keeps a record of the Earth's weather systems, one that it can guard long after its death, a death that may be caused by flood or earthquake, volcanic eruption or hurricane. It is *Porites*, the massive colonial coral that can slowly grow to 2 metres or more in diameter and dwarf the human diver approaching it, coring device in hand, in readiness to sample its life history.

Every now and again, El Niño arrives. The waters around the coral then become a little warmer, and a little fresher too as the rains fall harder. The coral keeps on growing, though with a little more difficulty, because it does not like the extra warmth (and some of its neighbours might die because of this). Still, as it builds its calcium carbonate skeleton, it incorporates a rather larger number of the light oxygen (^{16}O) atoms relative to the heavy (^{18}O) ones, partly because the higher temperature makes it favour that isotope, and partly because the rain is richer in this type of oxygen.[27]

The coral, as it lives and feeds, is quite unconscious of the cascade of effects spreading around the world as the warm water spreads eastwards past it, towards the west coast of the Americas. But it might, in its own way, register the change in the winds around it, winds that normally keep the warm water piled up far to the west in the Pacific, but now slacken—or even change direction.

As the winds slow, cold, nutrient-rich water is no longer dragged up from the ocean depths along the western American coastline. The surface waters warm, and the anchovies that usually teem in their billions disappear. The seabirds that normally gorge on them go hungry—and so too do the fishermen. Inland, it rains, often so hard that the hillsides give way amid floods and landslides. As the sediment-laden rivers pour into the sea, it is good news for some life-forms—scallops and shrimps and snails and octopi thrive.

The effects skitter around the world. Salmon in British Columbia decline—while those in Alaska boom. Californian sealion populations crash, but dolphin numbers pick up. Yet further, there is more rain than usual in central Africa—but severe drought in southern Africa. Our coral, now old and massive, may have been a youngster in 1789, when a particularly severe El Niño—it has been inferred—laid waste to harvests in Europe and helped spur the anger of the poor, an anger that would build high enough to remove the heads of the French aristocracy.

El Niño comes along every few years—while his little sister, La Niña, can visit in between times. Together they make up a kind of climate see-saw, called the Southern Oscillation (commonly abbreviated to ENSO, which a little confusingly stands for 'El Niño-Southern Oscillation') centred on the Pacific, but with changed weather patterns reverberating around the world, as in a colossal pinball machine.

La Niña, with all the perversity of a sibling, does the opposite to her brother. The westwards-blowing trade winds strengthen, more cold water is brought to the surface along the eastern Pacific, and the coastal regions of Chile and Peru suffer drought, while rains fall heavily across Malaysia and Indonesia. La Niña is also a turbulent child, shifting cyclone paths towards China and bringing more hurricanes to the Gulf States of the USA. And while she visits, the *Porites* corals carry on dreaming, and feeding—and lock a little more of the heavy oxygen isotope into their skeletons.

The ENSO phenomenon is, at bottom, a short-period climate cycle—though one whose origins and triggers remain a little mysterious. Over the last couple of centuries or so, human recording of its effects suggest that the El Niño part has come along at intervals of between two and seven years, though its strength and its effects can vary quite widely (La Niña is equally capricious). Understanding ENSO better would clearly be a good thing, given how profoundly it can impact upon human society. Not least, if climate is set to change, as we will discuss in the next chapter, will it make ENSO events fewer, or more common, or will there be no effect? To answer this we need to turn to the longer Pacific history—and to the patient corals, that year by year have collected data, regardless of whether or not there have been human observers to harvest and analyse them.

We can drill into living *Porites* corals, through the thin, living layer into the older skeleton beneath. This coral is particularly useful because it shows seasonal growth patterns that, in a cross-section of the entire colony, look remarkably similar to tree-rings.

One then needs to collect not just a living coral, but also an assortment of more ancient, dead examples, to splice their histories together, exactly as one does with tree-rings, except using uranium-thorium dating rather than carbon dating to place the dead corals approximately into a time framework. Then, within that framework, the bar-code-like patterns of good years and bad years are matched up to establish a precise history.

The pattern of ENSO events in the last millennium has been at least partly reconstructed in this fashion.[28] ENSO has varied in strength, it seems, and changes can occur quickly, on a timescale of decades. But, the patterns seem not to show any relation to the large features of this millennium, such as the Medieval Warm Period or the Little Ice Age. ENSO seems to dance to its own tune, rather than to any of the recognized external drivers, though there is some indication that very violent volcanic eruptions can trigger a strong El Niño event. On a longer timescale there are indications (from corals, tree-rings, and suchlike) that ENSO events were weak in the early part of the Holocene and during the last glacial phase, between 100,000 and 20,000 years ago, so perhaps astronomical factors do play a part at these long timescales.

These children may be very old—perhaps even as old as the Pacific. Their behaviour in the Pliocene, over three million years ago, has been an object of no little curiosity to scientists. If the warmth of the Pliocene represents something like our own near future, the way El Niño and La Niña behaved in the Pliocene may, perhaps, provide a clue to how they will behave over the next century or two.

The Pliocene children have proved elusive. Some researchers have proposed the idea that there were then permanent El Niño conditions,[29] because the slightly different tectonic arrangement of the Pacific land masses would have made it less easy for warm water to accumulate in the western Pacific, and because there is indeed some evidence that west-to-east temperature gradients across the Pacific were not as great as they are today. But analysis of fossil Pliocene *Porites* corals from the Pacific showed isotopic 'spikes' characteristic of an El Niño-La Niña variability rather like today's.[30] The debate, thus, seems interestingly poised.

After the Holocene

The world today is changing, very rapidly, and, somehow, that will impact upon this brother-and-sister partnership of El Niño and La Niña—and on much else. We might now speak of a global phenomenon, taking place all around us, that is different in type and scale and speed from anything that has yet taken place on Earth. A Neolithic hunter-gatherer, caught up from the depths of one of those endless early Holocene forests and dropped into the present, would find a world transformed. For better or worse, it is the new world of what many now call the Anthropocene.[31]

The Anthropocene Begins

Alternative futures

What is the future for the planet, and for climate? Gazing into crystal balls is a pastime that humans have a fascination for. It is also one in which they have a dismal record. A generation or two ago, there were predictions of cities made of glass or plastic, clothes of aluminium or asbestos, flying cars, the fall of nationalism and the rise of world government, the demise of religion, and robots taking over our tasks and ushering in an age of universal leisure for all. So much for all that.

When we move, then, to the almost limitless complexities and intersecting feedbacks of Earth's climate system, one might be forgiven for throwing in the towel straight away. This is a system, we must eternally remember, of which we have only partial understanding, even as we see today's weather patterns spin off from it. Go back into the deep past, and that climate and those long-vanished weather patterns leave only traces in strata that are, in large part, invisible to the naked eye. And of the future, of course, we have no samples, no deep boreholes, no fossils: the canvas is blank—indeed, as yet there is no canvas at all.

Yet, from those ancient stratal traces we can construct a picture of events that is both vivid and (within our levels of uncertainty) true. There is no reasonable doubt that 20,000 years ago massive ice sheets spread out from the poles—or, that 125,000 years ago there was a climate on Earth as temperate, within a degree or so, as the one we enjoy today. So, there are patterns, real patterns that we can use as guides to help us try, with the utmost caution and scepticism, to create pictures, scenarios, sketches of the climate of the future.

One might imagine alternative futures—or create them—particularly with the help of those elapsed realities. For instance, one might take that striking five-million year slice of climate history put together by Lorraine Lisiecki and Maureen Raymo (see Ch. 8). This reconstruction shows a steady and inexorable increase in the intensity of the astronomically driven highs and lows of climate, from the relatively warm 40,000-year world of the Pliocene Epoch, to the colder 40,000-year world of the early Quaternary, with its more clearly marked climate lows, to the dramatic 100,000-year world of the later Quaternary, with its intense swings of both cold and warmth. It is like looking at the sensors of some complicated and dangerous machine (which indeed the Earth is) as they begin to flash red and sound the alarm. The world's climate system has steadily become more extreme. So what is going on—and where is this leading?

This is a question that the climate scientists Tom Crowley and William Hyde asked a year or two back. They constructed a computer model of the Earth involving values ('boundary conditions') of parameters such as temperature, radiation balance, ice volume, carbon dioxide levels: a model that could replicate that pattern of steadily increasing climate variability. They ran it over three million years up to the present day—and then from there ran it forwards in time.[1]

The results were striking. The modelled future Earth, in the geologically near future of some 100,000 years hence, makes a sudden transition to a new, stable state—one of a much deeper and long-lasting age of ice. The authors call it a climate bifurcation point, and its appearance is compared with the way that financial markets can become more and more volatile—until they crash and reconfigure themselves.

In this newly created, computer-simulated world a new symmetry of the hemispheres has appeared, with ice sheets growing on Europe and Russia to rival those on Antarctica. The level of the modelled ocean plummets, as the ice grows, to levels some 300 metres lower than today. It is a vision of a world more ice-bound than at any time since the days of Snowball Earth, back in the Precambrian. It is a sober piece of science that has out-hyped the visions of the moguls of Hollywood. And yet it is a model, we must remember—an intelligent but imperfect projection, given the programmed conditions, of what *might* have happened in a world left to its own devices.

But the world has not been left alone, and, as the authors of the model noted, this is now a potential ice-world averted, a possible future indefinitely postponed. The injection in a couple of centuries of half a trillion tons of carbon from its deep-Earth stores into the atmosphere has been more than enough to forestall a deep freeze, even one a hundred thousand years in the future.

For we are now moving into a world where the control levers of climate are increasingly in the hands of humans. It is a world that has been termed the Anthropocene by Paul Crutzen,[2] who won a Nobel Prize in 1995 for unravelling mechanisms that damage the ozone layer. The changes go beyond climate, of course. They involve the landscape, biodiversity, ocean chemistry, and much else. But climate is such an important driver of so many aspects of the Earth system—of rates of erosion, of sea level, of the distribution (via species-specific tolerances of temperature and humidity) of animals and plants—that its history can fairly be said to be a central factor in Earth history.

And thus, climate will probably not unroll as if modern humans had not evolved and mightily prospered. Its course will unroll *differently*—and one possible outcome is that there will soon come another, but different climate bifurcation. We will consider the nature of this altered future scenario now. It may be a question that is doomed to obsolescence at the outset, for any future paths that might be indicated now will certainly be trumped by the real future that will unroll in ten, a hundred, a thousand years (and more) from now. Nevertheless it remains a valid question, and an important one, and, most of all, it is one that it is in our natures to ask—for we have a drive to link the past with the future via the ever-changing moment that is *now*. So here goes.

As a starting point, one might consider just how far back in time humans might have started to change the course of climate.

The early Anthropocene hypothesis

Our human species appeared about 160,000 years ago—not particularly long as animal species go. But some of our 'human' activities, such as the use of tools, evolved much earlier: *Homo habilis* used stone tools about two and a half million years ago, while *Australopithecus afarensis* (famously represented by the fossilized remains of a young female known as 'Lucy'), used sharp-edged stones to butcher meat three-quarters of a million years before that. This was simply, though, an improvised cultural means of making up for our pathetically diminutive claws and teeth. Our kind lived in the wild, but did not tame it. Technological evolution proceeded at glacial speed.

Even when our own species appeared, it lived in a similar fashion for more than nine-tenths of its duration. It became more adept at killing and eating other animals, to be sure, and the tools became more sophisticated. Around 10,000 years ago, more or less coincident with the last retreat of ice, about half the world's large mammal species disappeared. Did climate change kill them, or was it the growing numbers of increasingly skilful humans? The balance of opinion suggests that our species probably had a large and hairy hand, at least, in this mass kill.[3]

And this inferred orgy of predation has been suggested as the first time that humans began to influence world climate. Mammoths (and other ruminants) are prodigious eaters of plants—and are also prodigiously flatulent. The breakdown of plant matter in their stomachs produces large amounts of methane, which then emerges as best it can. Kill a lot of the world's methane producers, it has been suggested, and you may significantly affect global methane production.[4] The Greenland ice core shows a marked dip in atmospheric methane levels that coincides pretty well exactly with the thousand-year chill of the Younger Dryas event, thirteen thousand years ago, at the very end of the Pleistocene. What is more, this is just preceded, in the North American archaeological record, by the appearance of the sophisticated tools of the human 'Clovis' culture implicated in these large mammals' demise. Thus, here is yet another (see Ch. 9) possible mechanism to trigger that abrupt climate reversal of the Younger Dryas.

This intriguing idea of humans as climate modifiers (it is a little grand, yet, to call it a fully fledged hypothesis) has been mooted in the last couple of years. But for a more mature and well-founded (albeit still highly controversial) hypothesis of early human influence on climate, we need to look to the beginning of a more settled lifestyle. About eight thousand years ago came the beginnings of settled agriculture. Humans began burning and clearing the forests of the Middle East and Eurasia, in the first spread of agriculture, converting those carbon stores into carbon dioxide released into the air. Then, some five thousand years ago, fields began to be purposely flooded in Asia, to aid the growing of rice. The practice spread, and from those burgeoning artificial wetlands, large amounts of methane began to be emitted. As more and more of these greenhouse gases began to be released, the Earth's climate responded—and was warmed just sufficiently to halt the slide back into glacial conditions. In this view, we have the Neolithic farmers to thank for postponing the next advance of ice.

The climate scientist Bill Ruddiman of Virginia University has been developing this hypothesis now for more than a decade.[5] A central framework was provided by history (or, more precisely, stratigraphy), that first long detailed history of temperate and carbon dioxide levels provided by the Antarctic Vostok ice-core. In the three previous interglacials of the Vostok core, carbon dioxide and methane rise quickly to a high level and then slowly decline through the rest of the interglacial. In the Holocene, by contrast, they followed this pattern early on—but, a few thousand years later, began to rise again, at the time that the archaeological record shows the spread of farming. Carbon dioxide levels rose, slowly, from about 260 to 280 ppmv, and methane levels from 600 to 750 ppbv. The Holocene history of atmosphere and climate, said Ruddiman, thus seems quite distinct—and the difference is most plausibly the human factor. He has calculated that the extra greenhouse gases may have raised global temperatures by as much as 0.8 °C.

It is an elegant and persuasive idea, thoroughly worked up by a widely respected scientist with a good track record of bringing formerly way-out ideas into the mainstream (early in his career, he suggested that the rise of the Himalayas some forty million years ago would have altered the climate of Asia, an idea elaborated on by Maureen Raymo and now broadly regarded as justified). From an agro-archaeological point of view this seems plausible, as even with the much smaller human populations then, those early farmers with their slash-and-burn activities would have used large areas of land extensively, rather than smaller areas intensively, to grow their crops.[6] The carbon footprint, meal for meal, would have been high.

Nevertheless, is this hypothesis of 'early Anthropocene climate change' probable?

As more evidence came in, the story has become, at the least, more complicated. Firstly, as we have seen, the Vostok core did not sample the fourth interglacial before the present one. The subsequent, EPICA, core did this, though, as it plumbed over 800,000 years' worth of ice. That interglacial, the one labelled 'Stage 11', as we have seen (Ch. 9), is longer, and shows a more Holocene-like pattern of a late, slight rise of greenhouse gas levels and associated warming. Thus, the Holocene might naturally have more than fifteen thousand years to run, rendering unnecessary our 'rescue' by Neolithic farmers from the imminent return of ice.

A different angle was examined by Thomas Stocker, another very well-respected scientist who, like many of his colleagues, had been highly intrigued by Ruddiman's hypothesis when he first heard it. But even well-founded and likeable ideas need testing. Stocker and his colleagues carefully analysed the carbon isotope composition of the Holocene greenhouse gases preserved in the bubbles in the Antarctic ice, up to Medieval times.[7] To obtain precise and reliable measurements of the ~1 per cent of heavy ^{13}C atoms in the sample needed extraordinary care. At these tiny levels, contamination from even the scrupulously clean laboratory equipment emerged as a major headache. But after a lot of work, reliable numbers eventually emerged.

The test centred on the origin of the extra early Holocene carbon—was it from land or sea? If from the land, it should have shown an increase in the light carbon ^{12}C, for this is preferentially taken up by the vegetation. If from the sea, no such preference should be shown.

The Holocene pattern emerged as two clear steps. First there was the slight decrease in carbon dioxide levels between eleven and seven thousand years ago, from about 265 to about 260 ppmv. As this took place, the carbon in the CO_2 became distinctly heavier, suggesting that light carbon was being removed from the air and taken up by land vegetation, spreading and blooming as the ice retreated.

Then came the slow upswing in carbon dioxide levels, that clear feature recognized by Ruddiman, to reach 280 ppmv by Medieval times. Here, though, there was no corresponding influx of light carbon, as might be expected if carbon from land vegetation was pouring back into the atmosphere as humans cleared the forests. The proportion of light to heavy carbon in the trapped gases of this age stayed almost the same (there was a slight increase in the light carbon, but not enough for the data to be consistent with the Ruddiman model).

The result did not agree, then, with the notion of early human influence of climate, of an early Anthropocene. Stocker and his colleagues suggested that the forest clearances had only a minor effect, and that most of the carbon dioxide came from the oceans, released by reactions involving carbon-containing minerals, and from the massive growth in coral reefs, where for each molecule of carbonate deposited in a coral skeleton, a molecule of carbon dioxide is released.[8]

A promising idea dead and buried? Not quite, for responses to these objections are possible. First, Ruddiman noted that if one lines up the beginning of the Holocene with the beginning of Stage 11, as was done in the EPICA study, then that makes these two stages have a curious astronomical contrast—the precession would have the earlier interglacial developing during a peak of summer insolation, and the present one during a minimum. But line them up so that the Holocene represents just the second part of Stage 11, and the pattern of increase and decrease in northern summer sunshine becomes near-identical.[9] If that latter correlation is the more valid one, then the Holocene is not halfway through its course, but near its natural end—and so in a state to be influenced by Neolithic farmers. And, while only a minority of carbon dioxide might be directly attributable to early deforestation, the methane from the early rice-fields might yet have had sufficient impact to tip the climate into mild warming.

The debate, therefore, continues—an example of the trickiness of establishing cause-and-effect in the climate debate, even within the very interglacial in which we live. But, over this first 98 per cent of the Holocene, we have still been in what has been, relatively speaking, very great stability of atmospheric composition and of climate. The last 2 per cent of it has been extraordinary—unique, probably in Earth history—as regards what has happened to the atmosphere and to the Earth's surface. We wait, and wonder, how global climate will respond.

The carbon transfer

The maintenance of life, and the pursuit of prosperity, has been dependent upon energy of one form or another. For most of the Holocene, the energy source exploited by humans has been muscle-power (their own, or of captive animals—and sometimes of captive humans too). There was running water too, to power mill-wheels, and wind to turn windmills, and sunlight. Some of the sunlight was used directly, to convert shaped mud blocks into building bricks, for instance. But, much of that exploited sunlight was second-hand, captured by trees that were then cut down and burnt. And it was in this way—perhaps—that carbon dioxide levels crept up by a few parts per million over five thousand years or more.

But, since before Roman times, people had noticed that there was a type of rock that burned. This, too, could be exploited. People had to learn how to get *at* that material, mind, for the underground seams are capricious in their distribution. From Roman to medieval times, the techniques remained primitive, miners digging out what they could from surface exposures, or, to pursue the coal further, digging 'bell-pits' into the rock strata, short vertical shafts that widened outwards once the coal was encountered.

Coal was a more effective energy source than wood, giving off much more heat, pound for pound. It could be transported and stored more easily, too. Initially it was used for heating homes. Then, when the great age of powered machines started, it was there and ready. Largely due to this new energy source, new possibilities opened up, of materials and machinery and goods. Coal production began to climb, and continued rising through the nineteenth century. Soon after the start of the twentieth century, it exceeded, globally, a billion tons annually (it is now about seven billion tons annually).

Then, a century after the Industrial Revolution, a new source of energy appeared. It was a rock that flowed. Petroleum oil, here and there, seeped from the Earth, though obtaining this stuff from the rare seeps at the surface was a slow and messy process. Demand for petroleum increased. In the USA, the entrepreneurially minded Samuel Kier, who had been selling crude oil as 'a cure for all ailments, animal and human', invented a means of distilling it. This produced kerosene (paraffin)—something that could be used instead of whale oil (a diminishing resource, then) to burn cleanly in lamps.

On 27 August 1859, the legendary 'Colonel' Edwin L. Drake and his associates assembled on some ground near Titusville, Pennsylvania. The Colonel wasn't a military man, but the honorific had been used in correspondence to him, and it stuck. He had worked for the railways, and so had free rail travel—a vital qualification that gave him a job prospecting for the rock oil.

Oil seeps had been reported in Titusville, and the Colonel went to look. There were just a few gallons of crude a day, but promising enough. They tried to sink a mine to get more—but that filled with water. Then came the idea to try a technique borrowed from salt producers—drilling.[10] After only a few metres, though, the hole kept caving in. Then Drake had the idea that would make an industry possible. To stop the caving-in, he drove an iron tube down the hole, within which the drill bit could turn unhindered. By this means, they slowly drilled on, at the rate of about a metre a day. But the rock was not yielding oil, and there were sundry mishaps, such as the drilling engine catching fire. The locals came to jeer at 'Drake's folly'. The Colonel, though, was a determined man.

At 'sixty-nine feet' down, the drill-bit hit a rock fracture. By the next morning, the hole had filled with oil. They scooped some out and stored it in a bath. By such means, in 1859, the United States produced 275 tons of oil. That was the start. Things developed from there, as we know.

Petroleum oil is a dream energy source. It is packed with energy, easily modifiable to different grades, easily transportable by pipeline, ship, or tanker, easily piped into one's personal transport machine. It is just so *convenient*. And for a century it has driven the world. One only has to stand at a busy street corner and see the rivers of automated steel pour by, and imagine the same thing happening on every road in every country all over the world, barely stopping for nightfall. It is a phenomenon, and there has to be a price, of some sort, to pay.

The exploitation of ancient carbon is as much a geological as a social and historical phenomenon, for it is dismantling and restructuring the Earth's long-term carbon cycle. Much of the coal that powered the Industrial Revolution in Europe, and is still central to powering the continuing industrial revolution in the USA, China, central Europe, and India, is rewriting an episode of global carbon history that is more than 300 million years old. In the Carboniferous, the burgeoning of terrestrial plant life (Ch. 3) happened to coincide with the construction and burial of one of the world's greatest-ever delta swamp systems, one that stretched half-way around the world. As thousands of generations of the fantastical Carboniferous trees lived, died, and were buried, levels of atmospheric carbon dioxide dropped, and the long-lived Permo-Carboniferous glaciation began. In our wholesale burning of the remains of those Carboniferous forests, we are reversing the course of that particular geochemical history.

Simultaneously, we are reversing others, from other geological eras and periods. The fabulous Saudi Arabian oil represents the burial of millions of generations of planktonic algae that fell onto a sea floor made stagnant and lifeless by the great rise in sea level in the rapid global warming event that caused the collapse of the enormous end-Ordovician ice sheet. That sea floor, in burying carbon through the succeeding early Silurian Period, helped regulate the atmosphere, until ice could build up again (if not to the same gigantic scale).

The oil in our cars comes also from the North Sea, the Gulf coast, Venezuela, Nigeria, Siberia. Natural gas, too, comes with both oil and coal. And, as the largest and easiest underground stores of petroleum are being exhausted, we search for, and find, other sources. There is the oil that is so sticky and viscous that it cannot be pumped out of its reservoirs—these are the enormous tar sands of Athabasca and elsewhere, perhaps equal to the Saudi Arabian oil reserves in size: but one can dig out the sands and steam-heat them to loosen the oil and make it more runny, so that it can then be pumped out. It costs a lot of energy to extract—but the demand for liquid oil is so great that this energy is being expended.

There is now a scramble for a new resource: shale gas. This is gas held not within the loose sponge of a porous sandstone, but crammed within the tiny pores of an impermeable organic-rich shale. To get at this costs energy, too—it involves drilling down into the shale, and then forcing liquid into the borehole at pressures high enough to create a dense network of fractures in the shale. (Sand grains have been added to the liquid, to jam into the fractures to hold them open.) The gas is then caught and collected as it seeps into the fractures. How big is this resource? No one knows. Shale is the commonest sedimentary rock on Earth.

As well as the burning of fossil fuels, carbon dioxide is released as land clearances and tree-felling continue, quite as fast as in Medieval times, and sometimes startlingly faster. In 1997, some of the thick peat deposits in Indonesia caught fire, as forest was cleared for palm oil plantations. The smoke trails spread across the Pacific, and could be seen from space. In that single event was released the equivalent of somewhere between 10 and 40 per cent of 'normal' annual human-caused carbon dioxide emissions.[11] Carbon dioxide is released, also, from our prodigious production of cement and concrete (around 5 per cent of total emissions). And we must not forget methane, that greenhouse gas some twenty-five times more powerful than carbon dioxide, albeit a good deal more short-lived, bubbling from the decaying forests drowned within reservoirs and from the guts of our cattle. Moreover, in changing the terrestrial landscape we are altering the albedo of the Earth—the amount of the Sun's light and heat that it absorbs, re-radiates, and reflects.

We are doing all this at once, geologically in the blink of an eye.

How to warm a planet

People have wondered for some time now about the possible consequences, of both the seizure by humans of the planet in general, and of the use of the atmosphere as a dumping ground for trace gases in particular. The dark, satanic mills of Blake's *Jerusalem* excited scientific enquiry as well as an emotional response.

The groundwork regarding carbon dioxide and other greenhouse gases had been laid, as we have seen (Ch. 1), by Joseph Fourier, Horace de Saussure, John Tyndall, and others. It was Svante Arrhenius (1859–1927), though, the Swedish scientist, who first clearly predicted human-made global warming through emissions of carbon dioxide.[12] His scientific pedigree was outstanding. His Nobel Prize was awarded for discoveries that were about as fundamental as it is possible to be—the ionic nature of solutions, and the concept of activation energy in chemical reactions. And his work on carbon dioxide was prescient, almost spookily so. In 1896 Arrhenius calculated, for instance, that to reduce global temperature to glacial levels would need a reduction in atmospheric carbon dioxide levels to between 55 and 62 per cent of its (then) level—which is remarkably close to what the Antarctic ice-cores show.

In writing of this, Arrhenius was notably generous to his Swedish colleague Arvid Högbom, who he quotes at length (almost four pages' worth), saying that his English-speaking audience would likely find Högbom's original Swedish text a touch difficult. Högbom certainly had a vivid way with ideas. How much carbon dioxide was there in the atmosphere, for instance? About as much as would make a layer of pure carbon a mere millimetre thick at the Earth's surface. Högbom also recognized that carbon dioxide was not only taken out of the atmosphere by plants (that might be buried to make coal), but by the action of dissolved carbon dioxide in rainwater (carbonic acid) reacting with silicate minerals on land surfaces. And, while this weathering would in general keep carbon dioxide levels in the atmosphere in equilibrium, the dynamic and fluctuating nature of the process could 'conduce to remarkable alterations of carbonic acid in the air'. So the changes in atmospheric carbon dioxide that Arrhenius postulated between glacial and interglacial times could, therefore, have been expected.

Arrhenius later suggested that human burning of coal could, over time, warm the Earth. He thought that this would be a good thing, preventing glaciation and making more favourable conditions to feed an ever-growing human population. His prediction was that it would take about three thousand years for humans to double atmospheric carbon dioxide levels. In this he had not reckoned on progress.

The carbon rise

Carbon dioxide is a *trace* gas in the atmosphere, as Arvid Högbom's vivid illustration showed. Moreover it is one that is continually in flux, being absorbed and emitted by various materials, both living and non-living. It was also, in the days of classical wet chemistry, a gas whose concentration was not easy to measure. So why should one attempt to carry out a succession of precise measurements of this gas, while being aware that first, it was difficult and time-consuming and second, the results might not mean very much once one obtained them?

Yet this was essentially the task that Charles Keeling (1928–2005) set for himself. In 1954, he was a freshly graduated chemist of quite unusual determination, who had persuaded the California Institute of Technology in Pasadena to hire him. More or less on a whim, he decided to measure carbon dioxide concentrations in the atmosphere. He pursued this whim—with determination—for the rest of his life, and, in producing what is probably the key diagram in all of climate science, changed our view of the world.

A little like John Tyndall a century before, he had to fashion the means to make these measurements. No one previously had thought to measure atmospheric carbon dioxide carefully and systematically enough to show any differences in this gas from time to time or place to place. Besides, it was generally thought that any fluctuations would largely reflect local influences such as vegetation, houses, factories, and so on. No matter: armed with glass jars, chemical solutions, and home-made pressure gauges, he set to work. The early experiments were beset with the kind of mishaps that can beset any apprentice scientist. The notebook with his first outdoor measurements, taken in a forest, was stolen by a deer. Chasing frantically after the animal, he eventually retrieved the data—imprinted with teeth marks but, mercifully, uneaten.

The initial measurements showed simply a change from day to night (as plants took up carbon dioxide for photosynthesis during the day and released it through respiration at night). Then, as the months passed, an annual change was seen, as the northern hemisphere has more land, and therefore more terrestrial vegetation, than the south: in the summer more carbon dioxide is taken up as plants grow and bloom, while in the winter both the leaves and the rate of photosynthesis fall.

For many scientists, that would have been discovery enough. But, determined as ever, Keeling carried on taking measurements to see how they would change from year to year. By now, he had started taking measurements in probably the best-mixed air on the planet, as far from local and human influence as it is possible to be—in the middle of the Pacific Ocean at the summit of Hawaii. The readings started at an annual average of 315 ppmv (with an approximate 7 ppmv annual range), but each successive year the mean value was 1 ppmv higher, and the rate of increase was remarkably steady.

And so the pattern has gone on, inexorably. There are slight variations, associated either with major perturbations such as the Indonesian fires or, more typically, with changes to the natural carbon dioxide transfers between atmosphere, land, and ocean. But there are no large variations or reversals, no sign that (say) individual volcanic eruptions, enormous as they are, are having any significant effect. Humans are in the driving seat. And, over the past couple of decades, with more humans collectively using more energy, the rate has increased from 1 ppmv to some 2 ppmv a year. Over that past three decades, this trend has gone from being an obscure scientific pattern to being a matter of widespread concern, if not alarm. It is agreed internationally (with assorted ifs and buts) that it would be highly desirable if this trend could be stopped, or at least slowed. Instead, it has accelerated, and as we write, is still accelerating.

The amount of carbon being released into the atmosphere by human activity now exceeds nine billion tons annually (translated

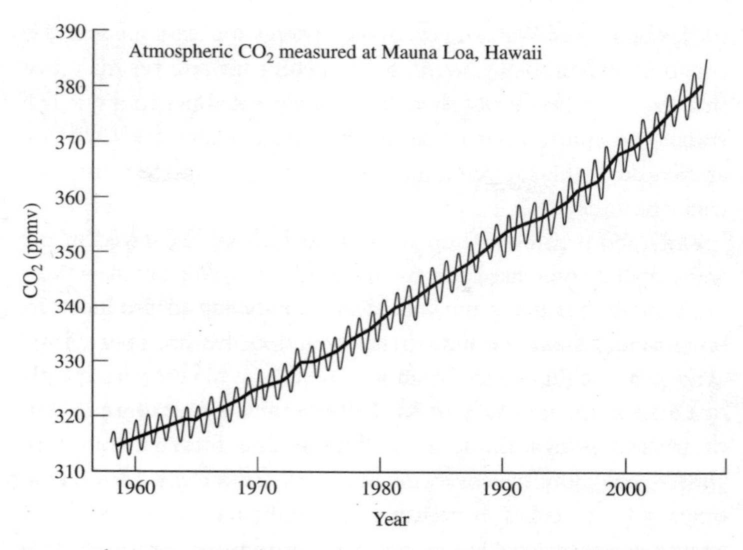

FIG. 27. The rise in atmospheric carbon dioxide over the last half-century, with its seasonal variations.

into carbon dioxide by adding the mass of the bound oxygen atoms, that becomes over thirty billion tons). Most of that comes from the burning of fossil fuels, which currently provide most of the world's energy. Each year's addition seems small by comparison with the Earth's carbon reservoirs. It is about a tenth of the amount annually exchanged by the opposed processes of photosynthesis and respiration, for example, or a hundredth of what is there in the atmosphere, or of that present in land vegetation.[13] Nevertheless, it is not an amount that can be lost or hidden in nature's carbon accounts. Year by year, it all adds up, as Charles Keeling's carbon dioxide curve shows.

The total amount of carbon released into the atmosphere by human activities over the past couple of centuries now adds up to some half a trillion (i.e. 500 billion) tons. Given present human numbers (at nearly seven billion, having more than doubled since the Second Word War) and ever-rising energy use, the trillionth ton of carbon will probably be emitted sometime towards the middle of this century, when atmospheric carbon dioxide concentrations will exceed 430 ppmv.[14] This is a *lot* faster than Arrhenius predicted but, as we said at the beginning of the chapter, reading the future is a tricky pastime.

Only about half this human-sourced carbon has stayed in the atmosphere. Some has been taken up by increased plant growth. A lot, though, has gone into the oceans, dissolving to produce carbonic acid.[15] Since the Industrial Revolution this has lowered the average pH of the oceans by about a tenth of a pH unit, from a pH of a little under 8.2 to about 8.1. This does not sound so much, but the pH scale is logarithmic, not arithmetic: thus, there are now some 30 per cent more hydrogen ions in the sea than in pre-industrial times.[16] By the end of this century, with humanity in the process of adding its second trillionth ton and atmospheric carbon dioxide likely to be over 700 ppmv (a level, by the way, not seen since the Eocene) then pH will have dropped by about 0.3–0.4 of a pH unit, and there will be about 100 per cent more hydrogen ions in the oceans.

This acidification of the oceans, often called 'the other carbon dioxide problem', is likely to be bad news, for nothing like it has occurred for many millions of years. The phenomenon threatens the many organisms in the sea that build skeletons of calcium carbonate, particularly those that employ the more soluble form, aragonite (such as all reef-building corals). Quite how bad it will be is currently a topic for intense research, because while predicting the pH of the oceans approaches being simple bucket-shop chemistry (so that once one knows the rate and amount of carbon dioxide emissions, the predictions are confident, and robust), the effects on the complex and varied biologies of different marine organisms are less easily predictable. Nevertheless, the studies so far suggest that as atmospheric carbon dioxide reaches somewhere in the range of

600 ppmv, coral reefs (for instance) will stop growing and start dissolving.[17]

The effect of this marine chemical revolution is all the greater because of its rapidity. The broadly comparable carbon release and ocean acidification event at the Paleocene–Eocene boundary[18] (Ch. 4) caused profound extinctions among bottom-living calcareous-shelled organisms, but had less effect at the surface. This suggests that the carbon was released to the atmosphere over thousands of years, allowing time for its dissolution products to be mixed into the oceans. By contrast, in this Anthropocene event, the bulk of the carbon release is taking place over two centuries or so. This is about a fifth of the normal 'ocean mixing time' so the surface-water regions, where the bulk of shelly marine organisms live, will be disproportionately harder hit.

By splicing Keeling's fifty-year carbon dioxide curve into the almost million-year-long carbon dioxide history obtained from the ice-cores, the scale of change becomes clear. Carbon dioxide levels are now at some 390 ppmv. This compares with a pre-industrial value of 280 ppmv (and, arguably, a pre-agrarian value of 260 ppmv)—a value that is similar, to within a couple of tens of ppmv, to those of interglacial phases extending back over that almost million-year span, while during glacial phases values dropped successively and almost metronomically to 180 ppmv. Thus, our collective activities have already elevated carbon dioxide above pre-industrial values by the same amount by which interglacial levels exceed glacial levels. Already, levels are probably similar to those of the Pliocene, three million years ago (Ch. 6), and it is likely that we will, in our children's lifetimes, exceed levels not seen since before the Antarctica ice sheet started growing, some thirty-three million years ago.

Methane levels, too, are now roughly double what they were in pre-industrial times. They rose rapidly up to 1990, then stabilized, and since 2007 they have been rising again. (The slowdown in the 1990s might have been due to effective plugging of many leaks from

Russian natural gas pipelines; the recent rise might, perhaps, be due to methane escaping from Russia's natural permafrost stores, as these begin to warm and melt.) Other, human-made greenhouse gases have appeared—some, such the chlorofluorocarbons, that were widely used in refrigerators, have exceedingly powerful warming effects (thousands of times that of carbon dioxide) as well as being damaging to the world's ozone layer. But, they are present in the atmosphere in minute levels, levels that are falling as a result of one of the world's few global environmental successes—the phasing out of these chemicals in the 1989 Montreal Protocol, to prevent further damage to the ozone layer.

As regards atmospheric and ocean chemistry, this is a blitzkrieg. What are—and will be—the effects on climate?

Climate now

The past century has seen a world that is increasingly, exponentially monitored and measured. Over the last hundred years, measurements have been taken of temperature, of rainfall, of wind, of ocean salinity and temperature. Not everywhere, not consistently: some parts of the world (the USA, Europe) are much better covered than others (Antarctica, say). And the gathering of routine data may be interrupted by war or, more prosaically, by budget crises. But, there is a history of basic climate parameters that, in broad outline and with error bars, is robust. Over the past few decades, the gathering of real-time data has been hugely expanded with the measurements taken by satellites in orbit, that allow us to see weather systems as they approach (and show meteorologists much more of the underlying detail). In the past few years, satellites have been launched specifically built to measure ice volume and, most recently, the details of carbon dioxide sources and sinks across the globe. With these latest forms of monitoring, the picture of what is happening will not be seen immediately, but will need to emerge through years (and decades) of consistent and patient measurement.

The picture that has emerged over the past century is clear. The early part of the century shows overall warming of the world (with decade-scale ups and downs), of some 0.3 °C. Then, between the 1940s and the 1970s, global temperature fell, overall, by about 0.2 °C. Then, starting about 1980, it rose sharply, by 0.6 °C, until the end of the century, following which the rise levelled off. This latest plateau may be at an end. As we write, 2010, just-elapsed, has tied with 2005 as the warmest year on record, and the thirty-fourth consecutive year hotter than the twentieth-century average. The globally averaged value, though, hides regional differences. Some parts of the world have stayed at much the same temperature, and some have cooled—such as the Atlantic region south of Greenland. Elsewhere, the temperature rises have been much higher; on the Antarctic Peninsula, average annual temperatures have risen by nearly 3 °C in the last half-century. This pattern of much greater changes at high latitudes is consistent with what we have seen when the world has warmed in the geological past (Chs. 4 and 6).

Is this warming necessarily human-made, a result of carbon emissions? Perhaps not. There is an inherent caution and scepticism in science, and rightly so. But, it is highly likely—much more likely than not—that the temperature trend is anthropogenic. At the end of the nineteenth century the world was emerging from the Little Ice Age, and, if this is part of one of those relatively subdued Holocene millennial-scale cycles, then a succeeding warming would be expected. Nevertheless, the early twentieth century was also part of the upward drive in the global economy, and in carbon emissions. But how, then, did global climate cool over the next several decades, even as the world's economy continued to grow?

Here the intersecting factors beyond greenhouse gases, both man-made and natural, come into play. Continuously ascending from factory chimneys and domestic hearths, and from burning forests too, has been smoke—fine particles of carbon, and also sulphur dioxide that combines with water in the atmosphere to form fine droplets.

The physics of atmospheric smoke and soot and sulphur (aerosols, they are termed), and the effect they have on both incoming light and outgoing infra-red radiation, is ferociously complex. But, on the whole, a filthier atmosphere seems to be a cooler one, at least in the lower levels of the atmosphere—a little like the way that global temperatures drop after a massive volcanic eruption sends fine ash into the stratosphere. And, humans, unlike volcanoes, have lately been emitting aerosols into the atmosphere without pause. Thus, one possibility is that atmospheric pollution masked the effect of raised carbon dioxide levels in the mid-twentieth century,[19] a masking effect that diminished as the dark, satanic mills were cleaned up in the last few decades (though they continue to emit, invisibly, just as much carbon dioxide). If this 'global dimming' is a real and important effect, then further steps to control visible pollution may raise temperatures yet higher.

Humans have brought about other changes, too, that have altered the radiative balance of the Earth's surface. The albedo, as we noted earlier, has changed substantially as natural vegetation has been cleared and replaced by agricultural land and by towns and cities, and as highly reflective ice has begun to shrink back to be replaced by more absorbent land and sea. The urban areas are domestically heated (and air-conditioned) heat islands, often several degrees higher than the surrounding countryside and these will add more complexity. The effects do not simply involve temperature, but other factors such as humidity. A rain forest to some extent makes its own weather, as vast amounts of moisture are continually recycled between forest canopy and sky via transpiration and rainfall. Cut enough of that forest down, and the transpiration stops, the

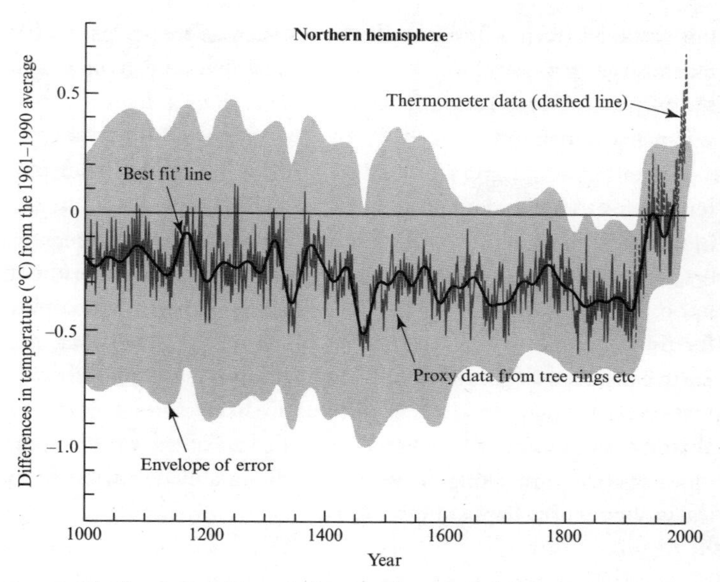

FIG. 28. The 'hockey stick' graph showing the rise in global temperatures over the last half-century.

rain washes off the landscape into the sea, and the whole area becomes drier.

The sharp modern rise in temperature—of some 0.7 °C in itself—over the twentieth century is striking: but how significant, in a longer context? In 1999, the scientist Michael Mann published a graph showing this late warming superimposed on the temperature trend of the last thousand years.[20] This 'hockey stick' graph showed the twentieth-century rise as exceeding in scale, and being more sharply defined, than that of the Medieval Warm Period. It was a controversial compilation, to put it mildly—not so much in the scientific community, but in the wider political arena. Much of the discussion (or rather, attack) focused around the error bars in the older part of the graph, where the temperature data is not instrumental, but gathered from a number of sources such as tree-rings, pollen assemblages preserved in lake sediments, and so on (Ch. 9). Could the Medieval Warm Period have been as much as a degree or two warmer still than today?

As matters stand, and after intense scrutiny, the hockey-stick pattern seems to be holding up reasonably well. Time will tell. We are in the midst of what one of the first of the modern climate researchers, Roger Revelle, called humanity's great geophysical experiment upon the Earth. The Earth's climate system is sufficiently complex for the outcome to be uncertain in much of the detail. But, the Earth's history is also sufficiently long for some reasonable consideration regarding what kinds of patterns are likely. Climate is not an abstract conception. It affects much of the rest of the world. What kinds of effects are taking place that can reasonably be attributed to the warming now happening?

Effects of warming

Climate is a good deal more than weather writ large. It drives much of what goes on at the Earth's surface. Where, first, is the heat? The extra half a degree of warming over the past half century warms the atmosphere and the land surface almost instantaneously. Elsewhere, though, it penetrates, slowly, deep into the waters of the ocean and into the rock of the ground.

The oceans have absorbed most of the extra heat that has been put into the Earth system over the last half-century, as the temperature of the upper several hundred metres has increased by a little more than half a degree.[21] The oceans have a high heat capacity, and the amount of this extra heat stored is, therefore, enormous. One study has put it at 14×10^{22} joules. That's just a number, but one joule is approximately the energy needed to lift a small apple a metre into the air, while 60 trillion (6×10^{13}) joules were released by the Hiroshima atom bomb. So the extra heat stored is the equivalent of a little more than 2×10^9 (billion) atom bombs.

That energy makes the water molecules move more energetically, to the extent that they move, on average, a little further apart. The whole ocean surface therefore lifts up, and this thermal expansion is responsible for about half the roughly 20-centimetre sea level rise of the past century (while the sea level has risen somewhat faster than that, at about 3 millimetres per year, since 1990). So far, the extra heat has spread into the top layers of water; it will be several centuries more before it spreads further down, and then is carried through the water mass by the deep ocean currents. The increased contrast between warmer, less dense surface layers and colder, denser bottom waters means that there is less interchange between them, and consequently the area of the nutrient-poor 'deserts' in the middle of the oceans has increased by a few million square kilometres.[22]

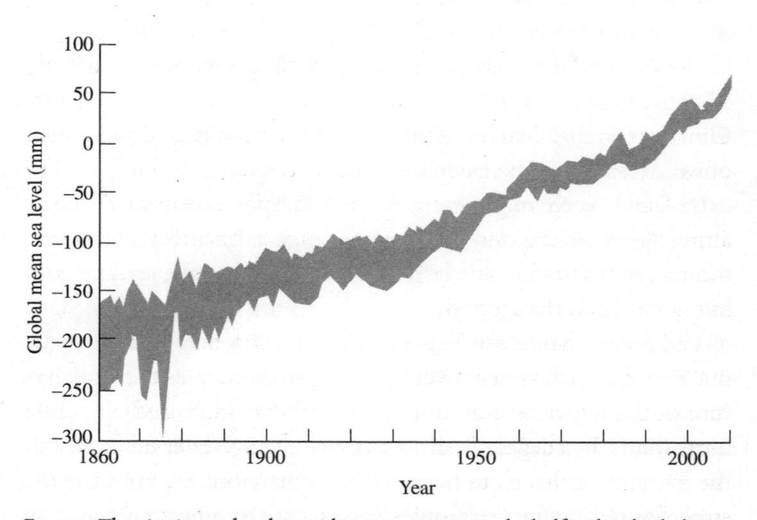

FIG. 29. The rise in sea level over the past century and a half—the shaded area reflects the degree of uncertainty, especially marked in the early years of this interval. The change is relative to the 1961–1990 average.

The extra warmth is felt, too, by the organisms that live in the oceans. Most of the attention of warming-related migrations has been to do with land animals and plants, but marine organisms feel the change more keenly. In large insulated water masses, temperature changes are lesser and slower than in the air, and so organisms have generally narrower temperature tolerances. Considerable range shifts have been observed on the land and in the sea, as animals and plants have migrated to keep within their habitability zones.[23]

For example, one of the dominant species in North Pacific waters is a diatom named *Neodenticula seminae*. A few years ago, this species spread into the Labrador Sea between Canada and Greenland, as the Arctic warmed, and it has since moved into Icelandic waters, reoccupying an area of ocean from which it had departed a million years ago, when the Quaternary ice ages intensified. In the Bering Sea, too, ecosystems are changing as the sea ice has shrunk back, with an increase in surface water fish, and decrease in sea floor dwellers (as the fish recycle nutrients that once reached the sea floor). On the other side of the world, in the seas around the Antarctic, warming is felt acutely by animals long adapted to a near-frozen world. There, starfish and molluscs that occupy the Antarctic shelf have nowhere to run to if the waters around them warm, except for a few refugia in the ocean deeps.

On land, such migrations, while less spectacular in absolute terms, have nevertheless been noted, especially at high latitudes, where temperatures have risen most. Timings of breeding and migration have changed. Alpine plants in the European Alps have moved higher, while tree lines have moved up generally. In Canada, the Red fox has moved northwards, while the Arctic fox has retreated. In Britain, bird species have been shifting their range northwards, by an average of a kilometre a year. Butterflies in North America and Europe have also shifted northwards: Edith's Checkerspot Butterfly, for instance, a species of the western USA, has moved over 90 kilometres northwards (and over 100 metres upslope) since the beginning of the twentieth century.

The world's animals and plants can thus be seen to be slowly on the move, after just over half a degree of temperature rise. What might happen as climate change really takes hold? One study suggested that with mid-range predictions of warming for the middle of the twenty-first century, somewhere between 15 and 37 per cent of species would be 'committed to extinction' from the warming effects alone.[24] They would have nowhere to run or to hide, especially in a world where ecosystems have been fragmented by agriculture and by building. And the world's organisms include those on which we rely for food. They are not in danger of extinction, but the threat of reduced crop yields over the coming decades, to feed a growing population, is very real.

As regards humans, what matters most is not so much average temperature, as the extremes of heat and cold (and of flood and drought and wind).[25] The past decade has seen record temperatures, and the impact on human society has been considerable—the 2003 heat wave in Europe caused over 14,000 deaths in France alone. It is impossible, though, to attribute blame unequivocally to climate change in the case of such events. But it is the extremes of temperature and weather that have profound human consequences (witness the effect of Hurricane Katrina on New Orleans, or the nature of the floods in Asia). Within a changing climate, it is the trend in the extremes—rather than in average temperature—that will probably be of greatest consequence. Among the consequences may be armed conflict—a common human reaction to increasing environmental difficulty, including climate perturbations.[26]

Other possible consequences are less easy to attribute. Are hurricanes, tornadoes, and cyclones more common than they used to be, because the world is a little warmer? This is not clear, though we have seen that the warmer Pliocene Pacific may have harboured powerful hurricanes (Ch. 6). There is more energy stored in the oceans that, simplistically, might drive more frequent and more damaging hurricanes. But, these delicately balanced spinning funnels of air depend on more than mere energy to grow and maintain them, and it has been suggested that the extra warmth might cause them to tend to tear themselves apart, rather than to help them to grow to catastrophic proportions. Increased spread of diseases such as malaria has frequently been quoted as something that will accompany global warming. Perhaps—but the spread of pathogens and their carriers is influenced by a complex range of factors, including the alacrity with which members of our species travel around the world, often with both mosquitoes and parasites in tow.

Global change

It is certain, therefore, that the Earth is changing—and in particular that it is changing as regards factors (greenhouse gas levels, albedo) that can influence the Earth's heat balance and therefore its climate. It is certain that, over the last century, the Earth has warmed already by something more than half a degree Celsius on average. It is judged highly likely by most scientists that this is due to human activities, largely to the (certainly human-made) increase in greenhouse gases being added to the atmosphere, an increase now taking place at a geologically unprecedented rate.

It seems likely, on current evidence, that the world will continue to warm, though the rate and the extent of this warming, in both the short and the long term, cannot be predicted precisely, even with today's sophisticated computer models—not least because we cannot predict quite how the carbon dioxide levels will continue to rise (the current rate of rise is near the high end of previous estimates). The world continues to be more sophisticated than the models we make of it. It is striking, though, that Aarhenius's predictions of how far global temperature would rise with a doubling of atmospheric carbon dioxide are not hugely dissimilar to successive predictions of the Intergovernmental Panel on Climate Change (IPCC).[27] Aarhenius, though, would have been hugely surprised at how quickly such a doubling could be achieved.

The degree to which global warming is a focus of controversy is a remarkable phenomenon in itself—and the venom of the attacks by 'climate change sceptics' or 'climate change deniers' on this branch of science is striking, and seems akin to the attacks on evolution by creationists. Why the venom? There is obviously a commercial angle, given how central hydrocarbons are to the economy, and how much money is made out of them. Any measures to reduce the use of hydrocarbons, or to make them more expensive, would reduce profits, or raise costs, in hydrocarbon-related industries. But not all the arguments are economic. Some people simply reject the notion that they—hardworking, honest taxpayers going about their lawful business—might be harming the Earth. Some consider the scientific community to be a cabal, conspiring in secret to fabricate a problem as a vehicle to obtain grant money and career advancement. Some consider scientists to be simply foolish, swept forwards on a wave of mass hysteria. Some wish to play devil's advocate, and by nature oppose established majority positions in any debate.

One cannot convince anyone whose mind is made up. And we, in writing this book, are well aware that we are geologists, and not climate modellers, or oceanographers, or statisticians. As geologists and more generally as scientists, we are fully paid-up sceptics too, by nature and by background, on any topic, and are always partial to devilish advocacy as a useful check on ideas. So, we do not hold to a *belief* that climate change will happen as per the forecasts of the IPCC. But, from a personal focus on the evidence gathered to date in fields that we can claim some expertise of, we *consider* it most likely—indeed consider it *very* likely—that climate has already started to change, towards greater global warmth.

Furthermore, it seems most likely that this change will continue—probably somewhat unsteadily—in decades, centuries, and millennia (and longer).

Key to this suspicion is the geohistorical context: first, that the climate over the past few million years—and particularly over the last half-million years or so—has been particularly febrile. It has undergone changes—often abrupt changes—every few thousand years (or more frequently). If such changes happened today, the effects would completely dislocate agriculture, the economy—in fact, human society itself. The Earth climate system is clearly in a long-term state in which it can turn on a sixpence (or a dime), and can react strongly and quickly to relatively modest stimuli.

Within this long-term phase of geohistorical turbulence, we as a human society have just happened to have found ourselves on the most stable climate plateau for at least four hundred thousand years. This precarious stability has allowed us as a species to expand tremendously, and to colonize and to grow our food on most of the habitable areas of the planet. Even within this interval of stability, the modest climate fluctuations that have occurred have proved catastrophic for some cultures.

In the last two hundred years, our numbers have expanded enormously; our energy use (and hence carbon emissions) have skyrocketed to the extent that the atmosphere now almost certainly holds more greenhouse gases than it has done since before the Quaternary; and there have been wholesale changes to the landscape and its reflective properties, and to the particulate content of the atmosphere. We are vigorously rocking a boat that has shown a marked tendency to capsize. The oceanographer Wallace Broecker has another analogy for the situation. 'Climate is an angry beast' he has said, adding that we are currently poking it with a stick. How angry might it get?

The future of the Anthropocene

Most of the widely publicized projections of climate change take us fifty or a hundred years into the future. That is a time appropriate for most current human and political concerns, allowing us to posit the predicted future against the future lives of our children and grandchildren. The consensus predictions of the IPCC over the coming century are now well-known: a 2–5 °C rise in average global temperature with a doubling of carbon dioxide from pre-industrial levels; an increase in extreme weather events; a change in the Earth's climate belts (such as an expansion of the arid/semi-arid zone); and a rise in sea level of the order of half a metre.

However, given the timescale of this book, that seems like an absurdly truncated forward look, one that does not look to where some sort of equilibrium might be arrived at. Neither does it involve the processes by which the Earth, many times, has seen its climate system perturbed from some previous norm, to reach some kind of high (or low) point, and then to move back to some sort of long-term equilibrium—which may not be the same as the pre-perturbation starting point. So, applying lessons from Earth history to the present situation—what long-term processes can one reasonably imagine?

The present rate of rise in carbon dioxide suggests that this is a process currently outside political control. Even though it is agreed by most governments that it is now quite urgent to control emissions, in practice this is not happening. Day-to-day politics and economic exigencies mean that too little is being done, too late, to alter the bottom line—the path of Keeling's carbon dioxide curve. It may be that some of the initiatives underway to discover new forms of energy (fusion power or effective photovoltaics, for instance) will bear fruit. But, even if something like this was discovered and spread through the world because it was better, simpler, cheaper than carbon-based power (like the spread of the computer chip or mobile phone), it would still take several decades to turn around the enormous ship, and rebuild the infrastructure, of the global economy. We discount carbon sequestration, because it is not happening and probably never will happen on a scale sufficient to make a difference (it would, though, be marvellous to be proved wrong on this).

So it seems that carbon dioxide levels will most likely double or triple, at least from pre-industrial values, to somewhere between 500 and 1000 ppmv within the next century or two (at these levels, by the way, the oceans may become biologically unrecognizable, as coral reefs largely disappear, as well as much of the calcium carbonate-shelled plankton). Levels of methane, already roughly double pre-agricultural levels, will likely have risen still further.

Before we consider the effects on climate, it is worth seeing how the atmosphere itself will evolve from this point, because the pattern of this, at least, is reasonably predictable, both from modelling and from the examples we have from the Earth's geological past.[28]

Carbon dioxide has a long residence time in the Earth's atmosphere—the average molecule stays there about a century (though this 'average' covers a wide variety of fates for individual molecules: many will be taken up by plants in a few years, while roughly a fifth of the total will still be in the atmosphere a millennium hence). Methane is much shorter-lived, at about ten years—but it typically disappears by oxidizing to carbon dioxide. So the simplest model is that carbon dioxide levels, having reached a peak, will very slowly decline, a small part perhaps being used up in extra plant growth, a proportion certainly being dissolved in the oceans, but much of the real regulation being done by the neutralization of the carbonic acid (carbon dioxide dissolved in rainwater) by reactions with the silicate rocks of the Earth's land surface. The final equilibrium is a level that is a little higher (all things being equal) than the original starting point. This is what happened in previous carbon outbursts and global warming events, such as in the Toarcian Age of the Jurassic and around the Paleocene–Eocene Thermal Maximum (PETM), the process taking somewhere between one and two hundred thousand years (Ch. 4). Thus, the recovery from the human disturbance of the atmosphere seems set to take a time roughly equivalent to the entire duration of *Homo sapiens* on Earth to date.

The warming of the Earth may well follow a similar course, with a PETM-like model of rapid warming to the order of 5–8 °C globally (more at the poles, less at the equator), global temperatures then declining over tens of thousands of years. That is the simplest model. But close examination of those ancient warming events shows that they were not simple, and that the course of the Anthropocene warming will probably be at least equally complex, because feedbacks, over geological timescales, will undoubtedly come into play.

One almost certain feedback is that warming of the oceans will put more water vapour into the atmosphere, amplifying itself (this is already built into predicted paths of warming such as those of the IPCC, although there are major uncertainties here because this question is inextricably linked with the future behaviour of those dynamic and complex structures, clouds). The alternating aridity and humidity of the glacial–interglacial cycles will have altered average global water vapour levels. But other feedbacks represent wild cards that, in some (unforeseeable) combination, will certainly be played.

The amounts of carbon stored in the vegetation and soils in different areas of the world will change, as climate belts shift between areas that are hot or cold, and wet or dry. It is unclear how this will affect the carbon balance of the world in sum—but this effect will certainly vary through time. The amount of carbon absorbed by plankton will change as the stratification of the oceans changes. The proportion of this carbon that sinks to the sea floor and is buried in sea floor sediments will also vary as oxygenation levels at the sea floor vary. The amount of dissolved carbon dioxide stored in the water mass of the oceans will vary, too. And then there is that overhanging cliff of carbon: that stored in methane hydrates.

Those ancient global warming events of the PETM and the Toarcian suggest that carbon release to the atmosphere was complex, and one plausible pattern is of modest initial carbon dioxide release (from volcanism?) and warming, that then triggered the release of methane hydrates from ocean floor sediments, which then caused the bulk of the warming. Today, there are substantial methane hydrates stored just below the sea floor, precariously poised between stability and decomposition. Luckily for us and for our immediate descendants, it will take several centuries at least for the increased warmth in the surface oceans to be carried to the depths of the oceans. But in the long term, that extra warmth will certainly get there. Rather more immediately, there is a large reservoir of methane hydrates stored in the permafrost regions of the Earth, in Siberia, Alaska, northern Canada. This is a store that was small or absent in the global greenhouse of the Mesozoic and early Tertiary, but it is one that is now vulnerable, and may well—in fact, probably will—release substantial amounts of methane over the coming decades to centuries.

Thus, in a long-term view, what humanity seems to be doing is not so much creating the entirety of a new global warming phenomenon, but rather lighting the blue touchpaper for an event that will acquire its own momentum, and that will then reverberate around the Earth according to its own internal dynamics. The course of this event cannot be predicted precisely, and is effectively irreversible, but it is likely, on past examples, to be neither brief nor small.

None of the analogues from the deep geological past are exact matches for what is happening now.[29] The recent glacial–interglacial changes involved very similar carbon transfers to the one which has been engineered by humanity so far, but those carbon dioxide variations were between glacial and interglacial state, and not between interglacial and the coming 'super-interglacial' state—and they were also much slower than what is happening now. So we are faced with a novelty. The duration of the atmospheric perturbation and the warming will be geologically novel, too. It will be of a similar scale to the 100,000-year Milankovitch orbital stretch cycles, and therefore will probably derail one or two of these[30] and all their attendant millennial Dansgaard–Oeschger cycles (and this is not considering the 'super-deep-freeze' future envisaged by Tom Crowley and William Hyde, as we discussed above). These glacial cycles will be replaced with something else that will probably have its own version of millennial-scale ups and downs, and that would also show the influence, at this scale, of the 20,000-year and 40,000-year precession and tilt cycles—much as the Toarcian event, at least, was modulated by these astronomical characters. The evolving Anthropocene—whether or not we will be around to help push it on—will not be simple.

Ocean waves

Of the consequences of climate change that appear to loom large in a geologist's mind, that of sea level comes high on the list. This is simply because fluctuations in sea level are, geologically, the norm (while Holocene-like stability, particularly in Ice Age times, is definitely the exception). Past sea level changes are also more or less easily visible in strata, and are used to help classify them; and, just as climate is a key driver of sea level, so sea level in turn drives many geological processes, of which just one is the movement of the Earth's coastlines.

Immediate concerns about sea level are not easily answered—that is, quite how fast will sea level rise by, say, the end of this century? Will it be somewhere between half a metre and a metre, if current rates (part thermal expansion, part ice melt) continue? This would be problematic, certainly, for low-lying areas: for Venice and Bangladesh, New Orleans and the Netherlands—but more or less manageable. Or, will it turn out to be double that or more as marginal ice sheets begin to collapse—a considerably more serious scenario, given how much of our coastal areas lie within a couple of metres of sea level. It is not just the increase in polar warmth that makes for longer summer melt seasons, and leads to warmer seas that nibble away at ice-sheet margins, that is the problem. Some of the smoke and soot we pour into the atmosphere lands on the ice, darkening it and making it absorb more of the Sun's energy.

In the longer term of centuries and millennia, though, it is highly probable that, as the world warms, ice will melt and sea level will rise.[31] If the global climate settles on a modest Pliocene-like climate (this is at the low end of the range of predictions, and as regards carbon dioxide levels we are at Pliocene levels already), then the sea level may eventually adjust to Pliocene-like levels, some 10–20 metres above those of today (see Ch. 6). This will be done by melting the vulnerable parts of the world's major ice sheets: those of west Antarctica and Greenland (the small mountain glaciers are insignificant in this respect). These ice masses will not melt steadily, but they will give way in a series of collapses, as ice-shelves break up and allow the ice behind to flow faster into the sea. Global sea level will thus rise in metre-scale jumps, just as it did in the many Heinrich events and catastrophic meltwater releases of the Pleistocene. This will proceed over centuries and millennia, giving rise to an Anthropocene geography quite different from the one that has persisted so unchangeably over so much of the Holocene. Whether we should be concerned about this longer-term prospect, of course, is a matter of debate—just as the people of Renaissance society may or may not have been concerned with how our own generation, now, might fare. We are now at least aware of the possible consequences for future generations.

Will the waters go on to rise higher, as succeeding millennia unfold? The next major step would be to reverse, temporarily or permanently, the step the world took across the Eocene–Oligocene boundary, as it grew the East Antarctica Ice Sheet. That would take us back into a Mesozoic-style greenhouse world, with sea levels up to 70 metres higher than today's. This type of world formed the bulk of Earth history, so it would be nothing new as such. The question is whether, with a global pattern of continents inferred to have encouraged glaciation, that is a reasonable possibility.

One might go further back in history for a possible parallel. During the end-Ordovician glaciation (Ch. 3) ice grew massively, as now, on a large southern continent (in that case on a conjoined South America and Africa). The brief dominance of ice was abruptly terminated by a very rapid and very large-scale deglaciation and dramatically rapid sea level rise, even while the overall tectonic configuration persisted. Ice did come back—but not for several million years and then not to the same extent.

Time will tell what on Earth will happen—a time that will stretch, probably, to beyond the human race (even if our collective span is to extend to a few million years, as have those of many species, past and present).

Climate of change?

What should we do?[32] It is clear that reducing carbon emissions is desirable, and necessary—even though we are not quite sure of the sensitivity of climate to carbon dioxide, and even though we do not know quite how climate change will unfold, or where the tipping points are, or even whether we have passed any significant tipping point. We must live with uncertainty, an uncertainty that will be ever-present, and be ever-changing through time too. But some things are more certain than others. To move aside from climate for a moment, it is worth every effort to reduce carbon emissions *simply* to prevent the oceans acidifying to the kind of levels that will kill off the coral reefs and much else. We have ten years left to save the oceans, the celebrated oceanographer Sylvia Earle said recently. That does not mean that catastrophe will overtake them this decade, but it does signify that if the Keeling carbon dioxide curve can't be budged off its upward course this decade, then our children and grandchildren will almost certainly inherit a hugely impoverished ocean—and one that is permanently impoverished, on any human scale.

Will the cutting of emissions by, say, a substantial carbon tax, wreck the economy? Well, it is unlikely to damage it to anything like the extent that, say, the global banking crisis has recently done. And it is unlikely to cost anything like carrying out a medium-sized war or two on foreign soil. The money thus raised might go in whole or in part into developing new energy sources—or even to sequestering carbon; or, it might, in whole or in part, be given back, in equal shares, as cash, to all citizens, as the NASA scientist James Hansen has suggested.

Some funding might go, perhaps, into the kind of geo-engineering schemes that people have suggested as climate fixes: mirrors in space, the pumping of sulphur dioxide into the atmosphere, and such. These are technological fixes of the last resort, and the science is still sufficiently exploratory to make it unclear whether or not the cure is worse than the disease. Better tread carefully.

It has been argued that climate has always changed, and will go on changing, and there is little that we can do to stop that. But that is a truism of such generality as to be useless. For sure, over long geological timescales climate has changed considerably, sometimes drastically. It has always regained equilibrium—eventually—and the Earth has not become uninhabitable. It really is the Goldilocks world, the only one of its type that we know, and probably something of a cosmic rarity. The question relevant to our current state is more specific.

Right now, we live in a particular climate state that, by and large, suits us. Any significant departure from that will make life harder and more dangerous for substantial sections of society—usually those who are poorest, and who have done least to cause the problem. So it makes sense to pay sufficient insurance to minimize the damage—and to help us adapt to the changes that will occur. We typically pay insurance (on houses, cars), to guard against slight risks of major damage. In considering global climate change, by contrast, there is generally agreed to be a large risk of very great damage. Those who say that there is no risk are simply wrong. And yet the insurance premiums remain unpaid.

Crazy, isn't it?

Notes

A BRIEF WORD ON TIME

1. A concise, authoritative resumé of the Geological Timescale is Ogg, Ogg, and Gradstein 2008.
2. A freely downloadable recent summary is Zalasiewicz et al. 2010

CHAPTER 1

1. Two nice essays that illustrate the debate about Martian water are Kerr (2008) and Bullock (2005).
2. Ingersoll (2007).
3. In some aspects of its chemistry the Moon is uncannily like Earth, suggesting that there was thorough mixing in the impact's aftermath: Halliday (2007).
4. Wilde et al. (2001).
5. The magnetic field was in place (if weaker than today) early in the Archaean, according to evidence of magnetized crystals in ancient rocks: Tarduno et al. (2010).
6. Modern-style plate tectonics probably started a little later, around three billion years ago, based upon changes (fascinatingly) in diamonds. See discussion by Van Kranendonk (2011).
7. Wacey (2010).
8. Fourier, though, was not the first to see this effect: the ability of material such as glass to allow light to pass through it but to be a barrier to heat—'*chaleur de feu*'—had been noted over a century earlier by another French savant, Edme Marriotte.
9. Fourier (1827).

10. Bunsen was a brilliant chemist and inspirational teacher, with a reputation for a singular focus on his work: on the day of his wedding (it is said) he did not turn up at the church. The guests found him locked in his laboratory. 'Go away!' he apparently shouted through the door, 'I am busy.' He never did marry.
11. Tyndall (1861).
12. His 'efficient seconders' in this instance were the Trustees of the British Museum: Mr Harlin, Mr Lettsom ('at the instance of Mr Darker'), Lady Murchison, Sir Emerson Tennant, Sir Philip Egerton, and Mr Pattison.
13. Rasmussen and Buick (1999).
14. Ohmototo, Watanabe, and Kumazawa (2004).
15. This is currently a point of deep interest among scientists working on Mars, where methane has been detected seeping from below that planet's surface.
16. Kasting (2010).
17. Newton (2010).
18. Goldblatt et al. (2009).
19. Rosing et al. (2010).
20. Robert and Chaussidon (2006).
21. Hren, Tice, and Chamberlain (2009).
22. Blake, Chang, and Lepland (2010).

CHAPTER 2

1. Young et al. (1998).
2. Polteau, Moore, and Tsikos (2006).
3. Kopp et al. (2005).
4. An eloquent—indeed, visionary—exponent of this early idea was Preston Cloud (1973).
5. Kappler et al. (2005).
6. Posth et al. (2008).
7. A good essay on the world's oxygenation is Kerr (2005).
8. See e.g. Bekker et al. (2004).
9. The strange world of the mid-Proterozoic is reviewed in Anbar and Knoll (2002).
10. His most accessible paper, for Western readers certainly, is probably Budyko (1969).
11. The work of Brian Harland and his colleague, Michael Hambrey, in collating evidence for this event, is detailed in their monumental *Earth's Pre-Pleistocene Glacial Record* (1981), and summarized in Hambrey and Harland (1985).

12. Donnadieu et al. (2004).
13. Nisbet and Nisbet (2008).
14. Hoffman et al. (1998); also Kirschvink (1992).
15. These alternative hypotheses are nicely and even-handedly discussed in Fairchild and Kennedy (2007).
16. Sohl and Chandler (2007).
17. The cap carbonates, in truth, remain controversial, whether abrupt and catastrophic or more prolonged: Kennedy and Christie-Blick (2011).
18. Allen and Hoffman (2005).
19. Planavsky et al. (2010); see also Filippelli (2010).

CHAPTER 3

1. Things ended more or less happily for William Smith—eventually. Towards the end of his life he received the recognition that was due him, medals from learned societies, and enough money to live in some comfort.
2. Elderfield (2010).
3. The shifting patterns of continents may in themselves have played a large part in climate change: Smith and Pickering (2003).
4. Landing and MacGhabann (2009).
5. Crowther and Rickards (1977).
6. Zalasiewicz et al. (2009).
7. Vandenbroucke et al. (2009). Similar patterns can be obtained using enigmatic fossilized egg-cases called chitinozoans: Vandenbroucke et al. (2010).
8. As an example: Davies et al. (2009).
9. e.g. Brenchley et al. (1994).
10. Quing et al. (1998). An intriguing development of this is that the spread of very early land plants—mosses—at this time accelerated rock weathering and so enhanced Co_2 drawdown (Lenton et al. 2012).
11. Page et al. (2007).
12. Falcon-Lang et al. (2009).
13. Wilman et al. (2004).
14. Montenegro et al. (2011).

CHAPTER 4

1. Greenwood, Basinger, and Smith (2010); see also Jahren and Sternberg (2008). For a fine account of the rich Eocene fossil forest of Axel Heiberg, that was then a little south of Ellesmere Island, at c.75° N, see Jahren (2007).
2. Eberle, Fricke, and Humphrey (2009); see also Paul (1988).

3. Brinkhuis et al. (2006).
4. The late Jurassic is indeed seen as a cool interval by Vaughan (2007) and Francis, Frakes, and Syktus (2005: See Further Reading).
5. Sellwood and Valdes (2007).
6. Pearson (2010).
7. Pyne (1980).
8. Hilgen (2010).
9. Gale et al. (1999).
10. Dean (2007).
11. The first documentation of global spread of black shales was by Hugh Jenkyns (1988); a good modern review is Cohen, Coe, and Kemp (2007).
12. Hesselbo et al. (2007).
13. Hesselbo et al. (2000).
14. McElwain, Wade-Murphy, and Hesselbo (2005).
15. Kemp et al. (2005).
16. Bowen and Zachos (2010). See also commentary on this in Archer (2010).
17. Cohen et al. (2004).
18. D. J. Thomas et al. (2002).
19. Crouch et al. (2001).
20. Zachos et al. (2005). See also Kump, Bralower, and Ridgewell (2009).
21. Sexton et al. (2011).
22. Pucéat, E. (2008), and for context see also Friedrich, Norris and Erbacher (2012).
23. Méhay et al. (2009).

CHAPTER 5

1. But not *the* first. Some sixty years before, the almost-forgotten James Eights, geologist, artist, and naturalist to the 'official' US expedition, in 1830, to the Antarctic region, had discovered fossil wood in the South Shetland Islands.
2. The story is in Garwood's splendid *Flat Earth* (2007).
3. Emiliani (1958).
4. A fine short biography is McCave and Elderfield (2011).
5. Shackleton (1967).
6. Zachos et al. (2001); an updated version is in Zachos, Dickens, and Zeebe (2008).
7. Coxall et al. (2005); Lear et al. (2008); and Wade et. al. 2012 .
8. Zanazzi et al. (2007).
9. Dupont-Nivet et al. (2007).
10. Merico, Tyrrell, and Wilson (2008).
11. Kennett (1977).

12. DeConto and Pollard (2003).
13. Pearson and Palmer (2000).
14. Beerling and Royer (2011).
15. Zachos et al. (1999).
16. Garzione (2008).
17. Reusch (2011).
18. Shevenell, Kennett, and Lea (2004).
19. Lyle et al. (2010).
20. Lewis et al. (2008); Williams et al. (2008).
21. Hill et al. (2007).

CHAPTER 6

1. Pagani et al. (2010).
2. Haywood et al. (2005).
3. Dowsett, H. J. (2007).
4. The shells, though, tell a more complicated story: Long and Zalasiewicz (2011).
5. Dowsett et al. (1992).
6. Kürschner et al. (1996).
7. Salzmann et al. (2011).
8. Salzmann et al. (2008).
9. Smellie et al. (2009).
10. Clark et al. (2010); Williams et al. (2010).
11. Naish et al. (2009).
12. Dowsett et al. (2010).

CHAPTER 7

1. Gibbard et al. (2010). Pillans and Naish (2004) summarize an earlier stage in the debate, and nicely summarize the environmental changes around the beginning of the Quaternary.
2. Willett, S. D. (2010).
3. Bartoli et al. (2005); Lunt et al. (2008).
4. Prueher and Rea (1995).
5. Haug et al. (2005); see also Billups (2005).

CHAPTER 8

1. Marvellously evoked in Rudwick's (2005) epic history.
2. Hays, Imbrie, and Shackleton (1976).

3. Lisiecki and Raymo (2005).
4. Bintanja and van der Wal (2008); see also Raymo and Huybers (2008).
5. Petit et al. (1999).
6. EPICA community members (2004). The commentary on this paper is good too: McManus (2004).
7. It has been estimated that the equivalent of the oceans' volume of water makes this underground journey every 70,000 years, picking up and releasing carbon on the way by a variety of chemical and biological processes.
8. Zimov, Schuur, and Chapin (2006); see also Ciais et al. (2012).
9. Jaccard et al. (2005).
10. Skinner et al. (2010).
11. Anderson and Carr (2010).
12. Dansgaard et al. (1993).
13. Bond et al. (1993).
14. Bond et al. (1992).
15. Bauch et al. (2003).
16. Blunier and Brook (2001). See also the excellent commentary by Shackleton (2001). A more recent study is EPICA community members (2006).
17. Ahn and Brook (2008).
18. Alley (2007).
19. MacAyeal (1993).
20. Braun et al. (2005).
21. Sun et al. (2006).
22. van der Hammen (1974); see also Hughen et al. (2004).
23. This is a step in the decay chain that will ultimately terminate in lead.
24. Wang et al. (2008). This has a good commentary: Overpeck and Cole (2008).
25. Finan (1999).
26. Wang et al. (2004). A marvellous new speleothem record from Peru—just 15 cm thick—shows that as Greenland warmed in D/O cycles and Northern hemisphere low latitudes became wetter—the Perusian Andes grew drier, and vice versa: Kanner et al. (2012), nicely commented on by Rodbell (2012).
27. Szabo, Haynes, and Maxwell (1995).

CHAPTER 9

1. The result, it has been suggested, of a climate suddenly akin to a 'nuclear winter', following the enormous eruption 70,000 years ago of the volcano Toba on the island of Sumatra at the eastern margin of the

Indian Ocean. The eruption was on a monstrous scale, it is certain—a thousand times greater than Mt. St Helens—but the evidence linking it to near-human demise remains slight: Balter (2010).

2. Placing a fix on absolute sea level is difficult, and controversial, because of the wildcard that land as well as sea can rise and fall, even on timescales of less than a million years (see also Ch. 6).
3. The evidence can be seen e.g. in ice, Alley (2000), and in terrestrial strata, Walker (2004).
4. Walker et al. (2008).
5. Steffensen et al. (2008). Commentary by Flückiger (2008).
6. Gould (1978); also Baker (2009).
7. Broecker (2006).
8. Murton et al. (2010).
9. Firestone et al. (2007).
10. Blanchon and Shaw (1995).
11. Rohling and Pälike (2005).
12. Played out on the margins of the southern North Sea e.g. there is Baeteman, Waller, and Kiden (2011) versus Behre (2007).
13. A fine array of this evidence is described in Mackay et al. (eds.) (2003).
14. <http://www.bbc.co.uk/news/science-environment-10958760>, accessed November 2011.
15. Gupta, Anderson, and Overpeck (2003).
16. Hu et al. (2008).
17. Sharpe (2002); see also Pearson and Betancourt (2002).
18. Chase et al. (2009).
19. Noren et al. (2002).
20. Wallace and Anderson (2010).
21. Fagan (2000) gives a vivid description of the impact of such change.
22. Foukal et al. (2006).
23. Eddy (1976).
24. Some can be a million times more energetic than anything a human-made particle accelerator can produce.
25. See Solanki et al. (2004) *v.* Muscheler et al. (2005). They disagree on how 'unusually' the Sun has behaved in recent years, but agree that its influence on recent temperature rise has been of little significance.
26. For instance, the current and unusually low minimum in the sunspot cycle (particularly within the ultraviolet part of the spectrum) has been linked with the cold and snowy winters in northern Europe and the US—

but to milder ones in southern Europe and Canada, and with little overall change in global average temperature: Ineson et al. (2011); see also discussion of this paper in Matthes (2011).

27. Ravelo (2006).
28. Wilson et al. (2010); see also Tudhope and Collins (2003).
29. Ravelo, Dekens, and McCarthy (2006).
30. Watanabe et al. (2011).
31. Crutzen (2002); see also Zalasiewicz et al. (2010).

CHAPTER 10

1. Crowley and Hyde (2008).
2. Crutzen (2002). The term first appeared in Crutzen and Stoermer (2000). Similar concepts, under various names, go back more than a century: see Steffen et al. (2011). Among them one might mention Revkin (1992).
3. Barnosky (2008).
4. Doughty, Wolf, and Field (2010).
5. Ruddiman (2003).
6. Ruddiman and Ellis (2009).
7. Elsig et al. (2009). See also commentary, Stocker (2009).
8. Because, in this process, two bicarbonate molecules combine—one carbon atom going into the carbonate precipitated in the coral skeleton, and one being released as a molecule of carbon dioxide.
9. Ruddiman (2007).
10. Salt, in those days before the use of the tin can, was fundamental to storing food—and those who controlled salt supplies could be as much captains of industry—and of empire—as the oil tycoons are today.
11. Page et al. (2002).
12. Arrhenius (1896).
13. And, as has been pointed out also, in a notorious newspaper article of 'climate denial' tone, it is 0.00022% of the amount emitted from the mantle in the Earth's history. That is (approximately) true, but *entirely* irrelevant.
14. Allen et al. (2009). There are several other thought-provoking articles in the same issue of *Nature*.
15. Doney et al. (2009).
16. In glacial times, with less carbon dioxide in the atmosphere, ocean pH was correspondingly higher, at about 8.3.
17. Kleypas and Yates (2009).
18. Kump, Bralower, and Ridgewell (2009).

19. Andreae, Jones, and Cox (2005).
20. Mann, Bradley, and Hughes (1999). The nature of the 20th century rise itself now seems clear: Witze (2012).
21. Hoegh-Guldberg and Bruno (2010).
22. Doney (2010); see also Hoegh-Guldberg and Bruno (2010).
23. Edwards (2009); see also Schofield et al. (2010).
24. C. D. Thomas et al. (2002).
25. Not just humans, of course. The recent 'bleaching' events affecting coral reefs have occurred during the marine equivalent of heatwaves, when high sea water temperatures triggered the expulsion of symbiotic algae from coral organisms, causing them to lose their bright colours and (usually) die.
26. Hsiang, Meng, and Cave (2011). See also commentary by Solow (2011).
27. IPCC (2007), available at <http://www.ipcc.ch/>, accessed November 2011.
28. Archer (2005); also Tyrrell (2011).
29. Haywood et al. (2011).
30. Tyrrell, Shepherd, and Castle (2007).
31. Rahmstorf (2007).
32. Tickell (2011); see also Steffen et al. (2011) and for a study of technological feasibility, Williams et al. (2012) and Shindell et al. (2012).

Further reading

General accounts

Alley, R. B. *The Two-Mile High Time Machine.*

Archer, D. 2009. *The Long Thaw: How Humans are Changing the Next 100,000 years of Earth's climate.* Princeton University Press.

Broecker, W. 2010. *The Great Ocean Conveyor: Discovering the Trigger for Abrupt Climate Change.* Princeton University Press.

Fagan, B. 2000. *The Little Ice Age: How Climate Made History 1300–1850.* Basic Books.

Imbrie, John and Imbrie, Katherine, 1986. *Ice Ages: Solving the Mystery,* Harvard University Press.

King, D. and Walker, G. 2009. *The Hot Topic: How to Tackle Global Warming and Still Keep the Lights On.* Bloomsbury Publishing.

Lovell, B. K. 2009. *Challenged by carbon: The oil industry and climate change.* Cambridge University Press.

Macdougall, D. 2004. *Frozen Earth: the once and future story of ice ages.* University of California Press.

Ruddiman, W. F. 2005. *Plows, Plagues and Petroleum: How Humans took Control of Climate.* Princeton University Press.

Turney, C. 2008. *Ice, Mud and Blood: Lessons from Climates Past.* MacMillan Science.

Walker, G. 2004. *Snowball Earth.* Bloomsbury Publishing.

Weiner, J. 1991. *The Next Hundred Years.* Bantam Press.

Textbooks/research volumes

Broecker, W. S. 1987. *How to build a habitable planet*. New York, Eldigio Press, 291 pp.

Cronin, T. M. 2010. *Palaeoclimates: Understanding Climate Change Past and Present*. Columbia University Press.

Ehlers, J. 1996. *Quaternary and Glacial Geology*. Wiley-Blackwell.

Florindo, F. and Siegert, M. (eds) 2009. *Antarctic climate evolution*. Elsevier.

Francis, J. E., Frakes, L. A. and Syktus, J. I. 2005. *Climate modes of the Phanerozoic*. Cambridge University Press.

Houghton, J. 2009 (4th ed.). *Global Warming: The Complete Briefing*. Cambridge University Press.

Huber, B. T., MacLeod, K.G. and Wing, S.L. (eds). 2000. *Warm climates in Earth history*. Cambridge University Press.

Letcher, T. P. (ed.) 2009. *Climate and global change: observed impacts on Planet Earth*. Elsevier.

Lowe, J. J. and Walker, M. J. C. 1997. *Reconstructing Quaternary Environments*. Prentice Hall.

Lunine, J.I. 1996. *Earth: Evolution of a Habitable World*. Cambridge University Press.

Roberts, N. 1998. *The Holocene: An Environmental History*. Wiley-Blackwell.

Williams, M., Haywood, A., Gregory, F. J. and Schmidt, D. N. (eds.). *Deep-time perspectives on climate change: marrying the signal from computer models and biological proxies*. The Micropalaeontological Society, Special Publication, The Geological Society, London.

Wilson, R. C. L., Drury, S. A. and Chapman, J. A. 1999. *The Great Ice Age*. Routledge.

References

Ahn, J., and Brook, E. J. (2008). Atmospheric CO_2 and climate on millennial time scales during the last glacial period. *Nature* **322**, 83–5.

Allen, P. A., and Hoffmann, P. F. (2005). Extreme winds and waves in the aftermath of a Neoproterozoic glaciation. *Nature* **433**, 123–7.

Allen, M. R., Frame, D. J., Hintingford, C., Jones, D., Lowe, J., Meinshausen, M., and Meinshuasen, N. (2009). Warming caused by cumulative carbon emissions towards the trillionth tonne. *Nature* **458**, 1163–6.

Alley, R. B. (2000). The Younger Dryas cold interval as viewed from central Greenland. *Quaternary Science Reviews* **19**, 213–26.

—— (2007). Wally was right: predictive ability of the North Atlantic 'conveyor belt' hypothesis for climate change. *Annual Review of Earth and Planetary Sciences* **35**, 241–72.

Anbar, A. D., and Knoll, A. H. (2002). Proterozoic Ocean chemistry and evolution: a bioinorganic bridge? *Science* **297**, 1137–42.

Anderson, R. F., and Carr, M-E. (2010). Uncorking the Southern Ocean's vintage CO_2. *Science* **328**, 1117–18.

Andreae, M. O., Jones, C. D., and Cox, P. M. (2005). Strong present-day aerosol cooling implies a hot future. *Nature* **435**, 1187–90.

Archer, D. (2005). Fate of Fossil Fuel in Geologic Time. *Journal of Geophysical Research* **110**, C09S05, doi: 10.1029/2004/2004JC002625.

—— (2010). How it went down last time. *Nature Geoscience* **3**, 819–20.

Arrhenius, S. (1896). On the influence of carbonic acid in the air upon the temperature of the ground. *Philosophical Magazine and Journal of Science*, series 5, **41**, 237–76.

Baeteman, C., Waller, M., and Kiden, K. (2011). Reconstructing middle to late Holocene sea-level change: A methodological review with particular reference to 'A new Holocene sea-level curve for the southern North Sea', presented by K.-E. Behre. *Boreas* 10.1111/j.1502-3885.2011.00207.x. ISSN 0300-9483 (published online); the latter also published in *Boreas* **36**, 82–102.

Baker, V. R. (2009). The Channelled Scabland: a retrospective. *Annual Review of Earth and Planetary Sciences* **37**, 393–411.

Balter, M. (2010). Of two minds about Toba's impact. *Science* **327**, 1187–8.

Barnosky, A. D. (2008). Megafauna biomass tradeoff as a driver of Quaternary and future extinctions. *PNAS* **105**, 11543–8.

Bartoli, G., Sarntheim, M., Weinelt, M., Erlenkeuser, H., Garbe-Schönberg, D., and Lea, D. W. (2005). Final closure of Panama and onset of northern hemisphere glaciation. *Earth and Planetary Science Letters* **237**, 33–44.

Bauch, D., Darling, K., Simstich, J., Bauch, H. A., Erlenkeuser, H., and Kroon, D. (2003). Palaeoceanographic implications of genetic variation in living *Neogloboquadrina pachyderma*. *Nature* **424**, 299–302.

Beerling, D. J., and Royer, D. L. (2011). Convergent CO_2 history. *Nature Geoscience* **4**, 418–20.

Behre, K.-E. (2007). A new Holocene sea-level curve for the southern North Sea. *Boreas* **36**, 82–102.

Bekker, A., Holland, H. D., Wang, P.-L., Rumble, D. III, Stein, H. J., Hannah, J. L., Coetzee, L. L., and Beukes, N. J. (2004). Dating the rise of atmospheric oxygen. *Nature* **427**, 117–20.

Billups, K. (2005). Snow maker for the ice ages. *Nature* **433**, 809–10.

Bintanja, R., and van der Wal, R. S. W. (2008). North American ice-sheet dynamics and the onset of the 100,000-year glacial cycles. *Nature* **454**, 869–72.

Blake, R. E., Chang, S. J., and Lepland, A. (2010). Phosphate oxygen isotope evidence for a temperate and biologically active Archaean ocean. *Nature* **464**, 1029–32.

Blanchon, P., and Shaw, J. (1995). Reef drowning during the last glaciation: Evidence for catastrophic sea-level rise and ice-sheet collapse. *Geology* **23**, 4–8.

Blunier, T., and Brook, E. J. (2001). Timing of millennial-scale climate change in Antarctica and Greenland during the last glacial period. *Nature* **291**, 109–11.

Bond, G., Broecker, W., Johnsen, S., McManus, J., Labeyrie, L., Jouzel, J., and Bonani, G. (1993). Correlations between North Atlantic sediments and Greenland ice. *Nature* **365**, 143–7.

—— Heinrich, H., and 12 others (1992). Evidence for massive discharges of icebergs into the North Atlantic ocean during the last glacial period. *Nature* **360**, 245–9.

Bowen, G. J., and Zachos, J. C. (2010). Rapid carbon sequestration at the termination of the Paleocene-Eocene Thermal Maximum. *Nature Geoscience* **3**, 866–9.

Braun, H., et al. (2005). Possible solar origin of the 1,470-year glacial climate cycle demonstrated in a coupled model. *Nature* **438**, 208–11.

Brenchley, P. J., Marshall, J. D., Carden, G. A. F., et al. (1994). Bathymetric and isotopic evidence for a short-lived Late Ordovician glaciation in a greenhouse period. *Geology* **22**, 295–8.

Brinkhuis, H., et al. (2006). Episodic fresh surface waters in the Eocene Arctic Ocean. *Nature* **441**, 606–10.

Broecker, W. S. (2006). Was the Younger Dryas triggered by a flood? *Science* **312**, 1146–7.

Budyko, M. I. (1969). The effects of solar radiation variations on the climate of the Earth. *Tellus* **21**, 611–19.

Bullock, M. A. (2005). The flow and ebb of water. Nature **438**, 1087–8.

Chase, B. M., Meadows, M. E., Scott, L., Thomas, D. S. G., Marais, E., Sealy, J., and Reimer, P. J. (2009). A record of rapid climate change preserved in hyrax middens from southwestern Africa. *Geology* **37**, 703–6.

Ciais, P. and 11 others. (2012). Large inert carbon pool in the terrestrial biosphere during the last Glacial Maximum. *Nature Geoscience* **5**, 74–9.

Clark, N., Williams, M., Okamura, B., Smellie, J., Nelson, A., Knowles, T., Taylor, P., Leng, M., Zalasiewicz, J. A., and Haywood, A. (2010). Early Pliocene Weddell Sea seasonality determined from bryozoans. *Stratigraphy* **7**, 199–206.

Cloud, P. (1973). Palaeoecological significance of the Banded Iron-Formation. *Economic Geology* **68**, 1135–43.

Cohen, A. S., Coe, A. L., and Kemp, D. B. (2007). The Late Palaeocene-Early Eocene and Toarcian (Early Jurassic) carbon isotope excursions: a comparison of their time scales, associated environmental changes, causes and consequences. *Journal of the Geological Society, London* **164**, 1093–108.

—— —— Harding, S. M., and Schwark, L. (2004). Osmium isotope evidence for the regulation of atmospheric CO2. *Geology* **32**, 157–60.

Coxall, H. K., Wilson, P. A., Pälike, H., Lear, C. H., and Backman, J. (2005). Rapid stepwise onset of Antarctic glaciation and deeper calcite compensation in the Pacific Ocean. *Nature* **433**, 53–7.

Crouch, E. M., Heilmann-Clausen, C., Brinkhuis, H., Morgans, H. E. G., Rogers, K. M., Egger, H., and Schmitz, B. (2001). Global dinoflagellate event associated with the late Paleocene thermal maximum. *Geology* **29**, 315–18.

Crowley, T. J., and Hyde, W. T. (2008), Transient nature of late Pleistocene climate variability. *Nature* **456**, 226–30.

Crowther, P. R., and Rickards, R. B. (1977). Cortical bandages and the graptolite zooid. *Geology and Palaeontology* **11**, 9–46.

Crutzen, P. J. (2002). Geology of Mankind. *Nature* **415**, 23.

—— and Stoermer, E. F. (2000). The 'Anthropocene'. *Global Change Newsletter* **41**, 17–18.

Dansgaard, W., Johnsen, S. L., Clausen, H. B., Dahk-Jensen, D., Gundestrup, N. S., Hammer, C. U., Hvidberg, C. S., Steffensen, J. P., Svelnbjörnsdottir, A. E., Jouzel, J., and Bond, D. (1993). Evidence for general instability of past climate from a 250-kyr ice-core record. *Nature* **364**, 218–20.

Davies, J. R., Waters, R. A., Williams, M., Wilson, D., Schofield, D. I., and Zalasiewicz, J. A. (2009). Sedimentary and faunal events revealed by a revised correlation of post-glacial Hirnantian (Late Ordovician) strata in the Welsh Basin, UK. *Geological Journal* **44**, 322–40.

Dean, W. T. (2007). Yorkshire jet and its links to Pliny the Elder. *Proceedings of the Yorkshire Geological Society* **56**, 261–5.

DeConto, R. M., and Pollard, D. (2003). Rapid Cenozoic glaciation of Antarctica induced by declining atmospheric CO2. *Nature* **421**, 245–9.

Doney, S. C. (2010). The growing human footprint on coastal and open-ocean biogeochemistry. *Science* **328**, 1512–16.

—— Fabry, V. J., Feely, R. A., and Kleypas, J. A. (2009). Ocean acidification: the other CO_2 problem. *Annual Review of Marine Science* **1**, 169–92.

Donnadieu, Y., Goddéris, Y., Ramstein, G., Nédélec, A., and Meert, J. (2004). A 'Snowball Earth' climate triggered by continental break-up through changes in run-off . *Nature* **428**, 304–6.

Doughty, C. E., Wolf, A., and Field, C. B. (2010). Biophysical feedbacks between the Pleistocene megafauna extinction and climate. *Geophysical Research Letters* **37**, L15703.

Dowsett, H. J. (2007). The PRISM palaeoclimate reconstruction and Pliocene sea-surface temperature, in M. Williams, A. Haywood, F. J. Gregory, and D. N. Schmidt (eds), *Deep-time Perspectives on Climate Change: Marrying the Signal from Computer Models and Biological Proxies* (London: The Geological Society for The Micropalaeontological Society), 459–80.

—— Cronin, T. M., Poore, R. Z., Thompson, R. S., Whatley, R. C., and Wood, A. M. (1992). Micropaleontological evidence for increased meridional heat transport in the North Atlantic Ocean during the Pliocene. *Science* **258**, 1133–5.

—— Robinson, M., Haywood, A., Salzmann, U., Hill, D., Sohl, L., Chandler, M., Williams, M., Foley, K., and Stoll, D. (2010). The PRISM 3D paleoenvironmental reconstruction. *Stratigraphy* **7**, 123–39.

Dupont-Nivet, G., Krijgsman, W., Langereis, C., Abels, H. A., Dai, S., and Fang, S. (2007). Tibetan plateau aridification linked to global cooling at the Eocene–Oligocene transition. *Nature* **445**, 635–8.

Eberle, J., Fricke, H., and Humphrey, J. (2009). Lower-latitude mammals as year-round residents in Eocene Arctic Forests. *Geology* **37**, 499–502.

Eddy, J. A. (1976). The Maunder Minimum. *Science* **192**, 1189–202.

Edwards, M. (2009). Sea life (pelagic and planktonic ecosystems) as an indicator of climate and global change, in T. P. Letcher (ed.), *Climate Change: Observed Impacts on Planet Earth* (Amsterdam: Elsevier), 233–51.

Elderfield, H. (2010). Seawater chemistry and climate. *Science* **327**, 1092–3.

Elsig, J., Schmitt, J., Leuenberger, D., Schneider, R., Eyer, M., Leuenberger, M., Joos, F., Fischer, H., and Stocker, T. F. (2009). Stable isotope constraints on Holocene carbon cycle changes from an Antarctic ice core. *Nature* **461**, 507–10.

Emiliani, C. (1958). Ancient temperatures. *Scientific American* **198**, 54–63.

EPICA community members (2004). Eight glacial cycles from an Antarctic ice core. *Nature* **429**, 623–8.

—— (2006). One-to-one coupling of glacial climate variability in Greenland and Antactica. *Nature* **444**, 195–8.

Fagan, Brian (2000), *The Little Ice Age: How Climate Made History, 1300–1850* (New York: Basic Books).

Fairchild, I. J., and Kennedy, M. J. (2007). Neoproterozoic glaciation in the Earth System. *Journal of the Geological Society of London* **164**, 895–921.

Falcon-Lang, H., DiMichele, W. A., Elrick, S., and Nelson, W. J. (2009). Going underground: in search of Carboniferous coal forests. *Geology Today* **25**, 181–4.

Filippelli, G. L. (2010). Phosphorus and the gust of fresh air. *Nature* **467**, 1052–3.

Finan, T. (1999). Drought and demagoguery: a political ecology of climate variability in NE Brazil. Paper from workshop on 'Public Philosophy, Environment and Social Justice', Carnegie Council, New York.

Firestone, R. B., West, A., Kennet, J. P., and 23 others. (2007). Evidence for an extraterrestrial impact 12,900 years ago that contributed to the megafaunal extinctions and the Younger Dryas cooling. *Proceedings of the National Academy of Science* **104**, 16016–21.

Flückiger, J. (2008). Did you say 'fast'? *Science* **321**, 650–1.

Foukal, P., Fröhlich, C., Spruit, H., and Wigley, T. M. L. (2006). Variations in solar luminosity and their effect on the Earth's climate. *Nature* **443**, 161–6.

Fourier, M. (1827). Mémoire sur les temperatures du globe terrestre et des espaces planetaires. *Mémoires de l'Académie Royale des Sciences de l'Institute de France* **7**, 570–604.

Friedrich, O., Norris, R. D. and Erbacher, J. 2012. Evolution of middle to Late Cretaceous oceans- **A55** m.y. record of Earth's temperature and carbon cycle. *Geology*, **40**, 107–110.

Gale, A. S., Young, J. R., Shackleton, N. J., Crowhurst, S. J., and Wray, D. S. (1999). Orbital tuning of Cenomanian marly chalk succession: towards a Milankovitch timescale for the Late Cretaceous. *Philosophical Transactions of the Royal Society of London* **A357**, 1825–9.

Garwood, C. (2007). *Flat Earth* (New York: Macmillan).

Garzione, C. M. (2008). Surface uplift of Tibet and global cooling. *Geology* **36**, 1003–4.

Gibbard, P. L., Head, M. J., Walker, M. J. C., and the members of the Subcommission on Quaternary Stratigraphy (2010). Formal ratification of the Quaternary System/Period and the Pleistocene Series/Epoch with a base at 2.58 Ma. *Journal of Quaternary Science* **25**, 96–102.

Goldblatt, C., Claire, M. W., Lenton, T. M., Matthews, A. J., Watson, A. J., and Zahnle, K. J. (2009). Nitrogen-enhanced greenhouse warming on Early Earth. *Nature Geoscience* **2**, 891–6.

Gould, S. J. (1978). The Great Scablands Debate. *Natural History* (August/September 1978), 12–18.

Greenwood, D. R., Basinger, J. F., and Smith, R. Y. (2010). How wet was the Arctic Eocene rain forest? Estimates of precipitation from the Paleogene Arctic macrofloras. *Geology* **38**, 15–18.

Gupta, A. K., Anderson, D. M., and Overpeck, J. T. (2003). Abrupt changes in the Asian monsoon during the Holocene and their links to the North Atlantic Ocean. *Nature* **421**, 354–6.

Halliday, A. N. (2007). Isotopic lunacy. *Nature* **450**, 356–7.

Hambrey, M. J., and Harland, W. B. (1985). The late Proterozoic glacial era. *Palaeogeography, Palaeoclimatology, Palaeocology* **51**, 255–72.

Harland, W. B., and Hambrey, M. J. (1981). *Earth's Pre-Pleistocene Glacial Record* (Cambridge: Cambridge University Press).

Haug, G. H., and 10 others (2005). North Pacific seasonality and the glaciation of North America 2.7 million years ago. *Nature* **433**, 821–5.

Hays, J. D., Imbrie, J., and Shackleton, N. J. (1976). Variations in the Earth's Orbit: Pacemaker of the Ice Ages. *Science* **194**, 1121–32.

Haywood, A. M., Dekens, P., Ravelo, A. C., and Williams, M. (2005). Warmer tropics during the mid-Pliocene? Evidence from alkenone paleothermometry and a fully coupled ocean-atmosphere GCM. *Geochemistry, Geophysics, Geosystems* **6**, 1–20.

—— Ridgwell, A., Lunt, D. J., Pound, D. J., Dowsett, H. J., Dolan, A. M., Francis, J. E., and Williams (2011). Are there pre-Quaternary analogues for future global warming? *Philosophical Transactions of the Royal Society* **A369**, 933–56.

Hesselbo, S. P., Gröcke, D. R., Jenkyns, H. C., Bjerrum, C. J., Farrimon, P., Morgans Bell, H. S. M., and Green, O. R. (2000). Massive dissociation of gas hydrate during a Jurassic oceanic anoxic event. *Nature* **406**, 392–5.

—— Jenkyns, H. C., Duarte, L. V., and Oliveira, L. C. V. (2007). Carbon-isotope record of the Early Jurassic (Toarcian) Ocean Anoxic Event from fossil wood and marine carbonate (Lusitanian Basin, Portugal). *Earth and Planetary Science Letters* **253**, 455–70.

Hilgen, F. J. (2010). Astronomical dating in the 19th century. *Earth-Science Reviews* **98**, 65–80.

Hill, D. J., Haywood, A. M., Hindmarsh, R. C. A., and Voldes, P. J. (2007). Characterising ice sheets during the mid Pliocene: evidence from data and models, in M. Williams, A. M. Haywood, J. Gregory, and D. Schmidt (eds) *Deep-time Perspectives on Climate Change: Marrying the Signal from Computer Models and Biological Proxies.* (The Micropalaeontological Society Special publication, The Geological Society, London) 517–38.

Hoegh-Guldberg, O., and Bruno, J. F. (2010). The impact of climate change on the world's marine ecosystems. *Science* **328**, 1523–8.

Hoffmann, P. F., Kaufman, A. J., Halverson, G. P., and Schrag, D. P. (1998). A Neoproterozoic Snowball Earth. *Science* **281**, 1342–6.

Hren, M. T., Tice, M. M., and Chamberlain, C. P. (2009). Oxygen and hydrogen isotope evidence for a temperate climate 3.42 billion years ago. *Nature* **462**, 205–8.

Hsiang, S. M., Meng, K. C., and Cave, M. A. (2011). Civil conflicts are associated with global climate. *Nature* **476**, 438–41.

Hu, C., Henderson, G. M., Huang, J., Xie, S., Sun, Y., and Johnson, K. R. (2008). Quantification of Holocene Asian monsoon rainfall from spatially separated cave records. *Earth and Planetary Science Letters* **266**, 221–32.

Hughen, K. A., Eglinton, T. L., Xu, L., and Makou, M. (2004). Abrupt tropical vegetation response to rapid climate changes. *Science* **304**, 1955–9.

Ineson, S., Scaife, A. A., Knight, J. R., Manners, J. C., Dunstone, N. J., Gray, L. J., and Haigh, J. D. (2011). Solar forcing of winter climate variability in the Northern Hemisphere. *Nature Geoscience* **4**, 753–7.

Ingersoll, A. P. (2007). Express dispatches. *Nature* **450**, 417–18.

IPCC (2007). Climate change 2007: synthesis report. Summary for policy makers: Available at: <http://www.ipcc.ch/>, accessed November 2011.

Jaccard, S. L., Haug, G. H., Sigman, D. M., Pedersen, T. F., Thierstein, H. R., and Röhl, U. (2005). Glacial/interglacial changes in Subarctic North Pacific stratification. *Science* **308**, 1003–6.

Jahren, A. H. (2007). The Arctic Forest of the Middle Eocene. *Annual Review of Earth and Planetary Sciences* **35**, 509–40.

—— and Sternberg, L. S. L. (2008). Annual patterns within the tree rings of the Arctic Middle Eocene (ca 45 Ma): isotopic signatures of precipitation, relative humidity and deciduousness. *Geology* **36**, 99–102.

Jenkyns, H. C. (1988). The Early Toarcian (Jurassic) Anoxic Event: Stratigraphic, sedimentary and geochemical evidence. *American Journal of Science* **288**, 101–51.

Kanner, L. C., Burns, S. J., Cheng, H. and Edwards, R. L. (2012). High-latitude forcing of the South American Summer monsoon during the Last Glacial. *Science* **335**, 570–73.

Kappler, A., Pasquero, C., Konhauser, K. O., and Newmann, D. K. (2005). Deposition of banded iron formations by anoxygenic phototrophic Fe(II)-oxidising bacteria. *Geology* **33**, 865–8.

Kasting, J. F. (2010). Faint young sun redux. *Nature* **464**, 687–9.

Kemp, D. B., Coe, A. L., Cohen, A. S., and Schwark, L. (2005). Astronomical pacing of methane release in the Early Jurassic Period. *Nature* **437**, 396–9.

Kennedy, M. J., and Christie-Blick, N. (2011). Condensation of cap carbonates during deglaciation. *Geology* **39**, 319–22.

Kennett, J. P. (1977). Cenozoic evolution of Antarctic glaciation, the circum-Antarctic oceans and their impact on global paleoceanography. *Journal of Geophysical Research* **82**, 3843–59.

Kerr, R. A. (2005). The Story of O_2. *Science* **308**, 1730–2.

—— (2008). Water everywhere on early Mars, but only for a geologic moment? *Science* **321**, 484–5.

Kirschvink, J. L. (1992). Late Proterozoic low-latitude glaciation, in J. W. Schopf (ed.), *The Proterozoic Biosphere* (Cambridge: Cambridge University Press), 51–2.

Kleypas, J. A., and Yates, K. K. (2009). Coral reefs and ocean acidification. *Oceanography* **22** (4), 108–17.

Kopp et al. (2005). The Palaeoproterozoic Snowball Earth: A climate disaster triggered by the evolution of oxygenic photosynthesis. *PNAS* **102**, 11131–6.

Kump, L., Bralower, T., and Ridgewell, A. (2009). Ocean acidification in deep time. *Oceanography* **22**, 94–107.

Kürschner, W. M., van der Burgh, J., Visscher, H., and Dilcher, D. J. (1996). Oak leaves as biosensors of late Neogene and early Pleistocene paleoatmospheric CO_2 concentrations. *Marine Micropalaeontology* **29**, 299–331.

Landing, E., and MacGhabann, B. A. (2009). First evidence for Cambrian glaciation provided by sections in Avalonian New Brunswick and Ireland: Additional data for Avalon-Gondwana separation by the earliest Palaeozoic. *Palaeogeography, Palaeoclimatology, Palaeoecology* **285**, 174–85.

Lear, C. H., Bailey, T. R., Pearson, P. N., Coxall, H. K., and Rosenthal, Y. (2008). Cooling and ice growth across the Eocene-Oligocene transition. *Geology* **36**, 251–4.

Lenton, T. M., Crouch, M., Johnson, M., Pires, N. and Dolan, L. (2012). First plants cooled the Ordovician. *Nature Geoscience* **5**, 86–89.

Lewis, A. R. and 12 others. Major Mid-Miocene cooling and extinction of tundra in continental Antarctica. *PNAS* **105**, 10676–80.

Lisiecki, L., and Raymo, M. E. (2005). A Pliocene-Pleistocene stack of 57 globally distributed benthic $\delta^{18}O$ records. *Paleoceanography* **20**, PA1003, doi: 10.1029/2004PA001071. 17 pp.

Long, P., and Zalasiewicz, J. A. (2011). The molluscan fauna of the Coralline Crag (Pliocene, Zanclean) at Raydon Hall, Suffolk: Palaeoecological significance reassessed. *Palaeogeography, Palaeoclimatology, Palaeoecology* **309**, 53–72.

Lunt, D. J., Valdes, P. J., Haywood, A., and Rutt, I. C. (2008). Closure of the Panama Seaway during the Pliocene: implications for climate and Northern Hemisphere glaciation. *Climate Dynamics* **30**, 1–18.

Lyle, M., Gibbs, S., Moore, T. C., and Rea, D. K. (2010). Late Oligocene initiation of the Antarctic Circumpolar Current: Evidence from the South Pacific. *Geology* **35**, 691–4.

MacAyeal, D. R. (1993). Binge/purge oscillations of the Laurentide ice-sheet as a cause of the North Atlantic's Heinrich events. *Paleoceanography* **8**, 775–84.

McCave, I. N., and Elderfield, H. (2011). Sir Nicholas John Shackleton 23 June 1937–24 January 2006. *Biographical Memoirs of the Royal Society*, doi: 10.1098/rsbm.2011.0005.

McElwain, J. C., Wade-Murphy, J., and Hesselbo, S. P. (2005). Changes in carbon dioxide during an oceanic anoxic event linked to intrusion into Gondwana coals. *Nature* **435**, 479–82.

Mackay, A., Battarbee, R., Birks, J., and Oldfield, F. (eds) (2003). *Global Change in the Holocene* (London: Hodder Arnold).

McManus, J. F. (2004). A great grand-daddy of ice cores. *Nature* 429, 611–12.

Mann, M. E., Bradley, R. S., and Hughes, M. K. (1999). Northern hemisphere temperatures during the last millennium: inferences, uncertainties and limitations. *Geophysical Research Letters* **26**, 759.

Matthes, K. (2011). Solar cycle and climate predictions. *Nature Geoscience* **4**, 735–6.

Méhay, S., Keller, C. E., Bernasconi, S. M., Weissert, H., Erba, E., Bottini, C., and Hoculi, P. A. (2009). A volcanic CO_2 pulse triggered the Cretaceous Oceanic Anoxic Event 1a and a biocalcification crisis. *Geology* **37**, 819–22.

Merico, A., Tyrrell, T., and Wilson, P. A. (2008). Eocene/Oligocene ocean de-acidification linked to Antarctic glaciation by sea-level fall. *Nature* **452**, 979–82.

Montenegro, A., Spence, P., Meissner, K. J., Eby, M., Melchin, M. J., and Johnston, S. T. (2011). Climate simulations of the Permian–Triassic boundary: ocean acidification and the extinction event. *Paleoceanography* **26**, doi:10.1029/2010PA002058.

Murton, J. B., Bateman, M. D., Dallimore, S. R., Teller, J. T., and Yang, Z. (2010). Identification of Younger Dryas outburst flood path from Lake Agassiz to the Arctic Ocean. *Nature* **464**, 740–3.

Muscheler, R., Joos, F., Müller, S. A., and Snowball, I. (2005). How unusual is today's solar activity? *Nature* **436**, E3–E4.

Naish, T., and 55 others (2009). Obliquity-paced Pliocene West Antarctic ice sheet oscillations. *Nature* **458**, 322–8.

Newton, A. (2010). Warming the early Earth. *Nature Geoscience* **3**, 458.

Nisbet, E. G., and Nisbet, R. E. R. (2008). Methane, oxygen, photosynthesis, rubisco and the regulation of air through time. *Philosophical Transactions of the Royal Society, London, B363*, 2745–54.

Noren, A. J., Bierman, P. R., Steig, E. J., Lini, A., and Southon, J. (2002). Millenial-scale storminess variability in the northeastern United States during the Holocene Epoch. *Nature* **419**, 821–4.

Ogg, J. G., Ogg, G., and Gradstein, F. M. (2008). *The Concise Geological Time Scale*. (Cambridge University Press), 184.

Ohmototo, H., Watanabe, Y., and Kumazawa, K. (2004). Evidence from massive siderite beds for a CO_2-rich atmosphere before ~1.8 billion years ago. *Nature* **429**, 395–9.

Overpeck, J., and Cole, J. (2008). The rhythm of the rains. *Nature* **451**, 1061–3.

Pagani, M., Liu, Z., LaRiviere, J., and Ravelo, A. C. (2010)). High Earth-system climate sensitivity determined from Pliocene carbon dioxide concentrations. *Nature Geoscience* **3**, 27–30.

Page, A. A., Zalasiewicz, J. A., Williams, M., and Popov, L. E. (2007). Were transgressive black shales a negative feedback modulating eustasy in the Early Palaeozoic Icehouse?, in M. Williams, A. M. Haywood, F. J. Gregory, and D. N. Schmidt (eds), *Deep-time Perspectives on Climate Change: Marrying the Signals from Computer Models and Biological Proxies* (London: The Geological Society for The Micropalaeontological Society), 123–56.

Page, S. E., Siegert, F., Rieley, J. O., Boehm, H-D. V., Jaya, A., and Limin, S. (2002). The amount of peat released from peat and forest fires during 1997. *Nature* **420**, 61–5.

Paul, G. L. (1988). Physiological, migratorial, climatological, geophysical, survival and evolutionary implications of Cretaceous polar dinosaurs. *Journal of Paleontology* **62**, 640–52.

Pearson, P. N. (2010). Increased atmospheric CO_2 during the Middle Eocene. *Science* **330**, 763–4.

—— and Palmer, M. (2000). Atmospheric carbon dioxide concentrations over the past 60 million years. *Nature* **406**, 695–9.

Pearson, S., and Betancourt, J. L. (2002). Understanding arid environments using fossil rodent middens. *Journal of Arid Environments* **50**, 499–511.

Petit, J. R., and 18 others. (1999). Climate and atmospheric history of the past 420,000 years from Vostock ice core, Antarctica. *Nature* **399**, 429–36.

Pillans, B., and Naish, T. (2004). Defining the Quaternary. *Quaternary Science Reviews* **23**, 2271–82.

Planavsky, N. J., Rouxel, O. I., Bekker, A., Lalonde, S. A., Konhauser, K. O., Reinhard, C. T., and Lyons, T. W. (2010). The evolution of the marine phaospate reservoir. *Nature* **467**, 1088–90.

Polteau, S., Moore, J. M., and Tsikos, H. (2006). The geology and geochemistry of the Palaeoproterozoic Makganyene tillite. *Precambrian Research* **148**, 257–74.

Posth, N. S., Hegler, F., Konhauser, K. O., and Kappler, K. (2008). Alternating Si and Fe deposition caused by temperature fluctuations in Precambrian oceans. *Nature Geoscience* **1**, 703–8.

Prueher, L. M., and Rea, D. K. (1995). Sudden onset of Northern Hemisphere glaciation—are volcanoes the smoking gun? *EOS*, **76**/46, F306.

Pucéat, E. (2008). A new breath of life for anoxia. *Geology* **36**, 831–2.

Pyne, S. J. (1980). *Grove Karl Gilbert—A Great Engine of Research* (Austin: University of Texas Press).

Quing, H., Barnes, C. R., Buhl, D., and Veizer, J. (1998). The strontium isotopic composition of Ordovician and Silurian brachiopods: relationships to geological events and implications for coeval seawater. *Geochimica et Cosmochimica Acta* **62**, 1721–33.

Rahmstorf, S. (2007). A semi-empirical approach to projecting future sea-level rise. *Science* **315**, 368–70.

Rasmussen, B., and Buick, R. (1999). Redox state of the Archaean atmosphere: evidence from detrital heavy minerals in ca 3250–2750 Ma sandstones from the Pilbara Craton, Australia. *Geology* **27**, 115–18.

Ravelo, A. C. (2006). Walker circulation and global warming: lessons from the geological past. *Oceanography* **19**/4, 114–22.
—— Dekens, P. S., and McCarthy, M. (2006). Evidence for El Niño-like conditions during the Pliocene. *GSA Today* **16**/3, 4–11.
Raymo, M., and Huybers, P. (2008). Unlocking the mysteries of the ice ages. *Nature* **451**, 284–5.
Reusch, D. N. (2011). New Caledonian carbon sinks at the onset of Antarctic glaciation. *Geology* **39**, 807–10.
Revkin, A. (1992). Anthrocene, in *Global Warming: Understanding the Forecast* (New York: Abbeville).
Robert, F., and Chaussidon, M. (2006). A palaeotemperature curve for the Precambrian oceans based on silicon isotopes in cherts. *Nature* **443**, 969–72.
Rodbell, D. R. (2012). Marching in near lock-step. *Science*, **335**, 548–9.
Rohling, E. J., and Pälike, H. (2005). Centennial-scale climate cooling with a sudden cold event around 8,200 years ago. *Nature* **434**, 975–8.
Rosing, M. T., Bird, D. K., Sleep, N. H., and Bjerrum, C. J. (2010). No climate paradox under the faint early sun. *Nature* **464**, 744–7.
Ruddiman, W. F. (2003). The anthropogenic Greenhouse Era began thousands of years ago. *Climate Change* **61**, 261–93.
—— (2007). The early anthropogenic hypothesis: challenges and responses. *Reviews of Geophysics* **45**, RG4001, 1–37.
—— and Ellis, E. C. (2009). Effect of per-capita land-use changes on Holocene forest clearance and CO2 emissions. *Quaternary Science Reviews* **28**, 3011–15.
Rudwick, M. (2005). The problem of erratic blocks (1810–14), in *Bursting the Limits of Time* (Chicago: University of Chicago Press), 571–85.
Salzmann, U., Haywood, A., Lunt, D., Valdes, P., and Hill, D. (2008). A new global biome reconstruction for the Middle Pliocene. *Global Ecology and Biogeography* **17**, 432–47.
—— Williams, M., Johnson, A. L. A., Kender, S., and Zalasiewicz, J. A. (2011). Climate and environment of a Pliocene warm world. *Palaeogeography, Palaeoclimatology, Palaeoecology* **309**, 1–8.
Schofield, O., Ducklow, H. W., Martinson, D. G., Meredith, M. P., Moline, M. A., and Fraser, W. R. (2010). How do polar marine ecosystems respond to rapid climate change? *Science* **328**, 1520–3.
Sellwood, B., and Valdes, P. (2007). Mesozoic climates, in M. Williams, A. Haywood, F. J. Gregory, and D. N. Schmidt (eds), *Deep-time Perspectives on Climate Change: Marrying the Signal from Computer Models and Biological Proxies* (London: The Geological Society for The Micropalaeontological Society), 201–24.

Sexton, P. F., Norris, R. D., Wilson, P. A., Pälike, H., Westerhold, T., Röhl, U., Bolton, C. T., and Gibbs, S. (2011). Eocene global warming events driven by ventilation of oceanic dissolved carbon. *Nature* **471**, 349–52.

Shackleton, N. J. (1967). Oxygen isotope analyses and Pleistocene temperature reassessed. *Nature* **215**, 15–17.

—— (2001). Climate change across the hemispheres. *Nature* **291**, 58–9.

Sharpe, S. E. (2002). Constructing seasonal climograph overlap envelopes from packrat midden contents, Dinosaur National Monument. *Quaternary Research* **57**, 306–13.

Shevenell, A. E., Kennett, J. P., and Lea, D. W. (2004). Middle Miocene Southern Ocean cooling and Antarctic cryosphere expansion. *Science* **305**, 1766–70.

Shindell, D. and 23 others (2012). Simultaneously mitigating near-term climate change and improving human health and security. *Science* **335**, 183–8.

Skinner L. C., Fallon, S., Waelbroeck, C., Michel, E., and Barker, S. (2010). Ventilation of the deep Southern Ocean and deglacial CO2 rise. *Science* **328**, 1147–51.

Smellie, J. L., Haywood, A. M., Hillenbrand, C-D., Lunt, D. J., and Valdes, P. J. (2009). Nature of the Antarctic Peninsula Ice Sheet during the Pliocene: geological evidence and modeling results compared. *Earth-Science Reviews* **94**, 79–94.

Smith, A. G., and Pickering, K. T. (2003). Oceanic gateways as a critical factor to initiate icehouse Earth. *Journal of the Geological Society of London* **160**, 337–40.

Sohl, L. E., and Chandler, M. A. (2007). Reconstructing Neoproterozoic palaeoclimates using a combined data/modeling approach, in M. Williams, A. M. Haywood, F. J. Gregory, and D. N. Schmidt (eds.), *Deep-time Perspectives on Climate Change: Marrying the Signals from Computer Models and Biological Proxies* (London: The Geological Society for The Micropalaeontological Society), 61–80.

Solanki, S. K., Usoskin, I. G., Kromer, B., Schüssler, M., and Beer, J. (2004). Unusual activity of the sun during recent decades compared to the last 11,000 years. *Nature* **431**, 1084–7.

Solow, A. R. (2011). Climate for conflict. *Nature* **476**, 406–7.

Steffen, W., Grinevald, J., Crutzen, P., and McNeill, J. (2011). The Anthropocene: conceptual and historical perspectives. *Philosophical Transactions of the Royal Society of London* **A369**, 842–67.

—— Persson Å., Deutsch, L., Zalasiewicz, J., Williams, M., Richardson, K., Crumley, C., Crutzen, P., Folke, C., Gordon, L., Molina, M., Ramanathan, V., Rockström, J., Scheffer, M., Schellnhuber, J., and Svedin, U. (2011). The Anthropocene: from global change to planetary stewardship. *Ambio.* **40**, 739–61.

Steffensen, J. P., and 19 others. (2008). High-resolution Greenland ice core data show abrupt climate change happens in a few years. *Science* **321**, 680–4.

Stocker, T. (2009). Making the paper. *Nature* **461**, 446.

Sun, Y., Clemens, S. C., An, Z., and Yu, Z. (2006). Astronomical timescale and palaeoclimatic implication of stacked 3.6-Myr monsoon records from the Chinese loess plateau. *Quaternary Science Reviews* **25**, 22–48.

Szabo, B. J., Haynes, C. V., Jr., and Maxwell, T. A. (1995). Ages of Quaternary pluvial episodes determined by uranium-series and radiocarbon dating of lacustrine deposits of eastern Sahara. *Palaeogeography, Palaeoclimatology, Palaeoecology* **113**, 227–42.

Tarduno, J. A. et al. (2010). Geodynamo, solar wind and magnetopause 3.4 to 3.45 billion years ago. *Science* **327**, 1238–40.

Thomas, C. D., and 18 others. (2002). Extinction risk from climate change. *Nature* **427**, 145–8.

Thomas, D. J., Zachos, J. C., Bralower, T. J., Thomas, E., and Bohaty, S. (2002). Warming the fuel for the fire: evidence for the thermal dissociation of methane hydrate during the Palaeocene-Eocene thermal maximum. *Geology* **30**, 1067–70.

Tickell, C. (2011). Societal responses to the Anthropocene. *Philosophical Transactions of the Royal Society* **A369**, 926–32.

Tudhope, S., and Collins, M. (2003). The past and future of El Niño. *Nature* **424**, 261–2.

Tyndall, J. (1861). On the absorption and radiation of heat by gases and vapours, and on the physical connection of radiation, absorption and conduction—the Bakerian Lecture. *London, Edinburgh and Dublin Philosophical Magazine and Journal of Science (Fourth Series)* **22** (146), 169–94, 273–85.

Tyrrell, T. (2011). Anthropogenic modification of the oceans. *Philosophical Transactions of the Royal Society* **A369**, 887–908.

—— Shepherd, J. G., and Castle, S. (2007). The long-term legacy of fossil fuels. *Tellus B* **59**, 664–72.

Vandenbroucke, T. R. A., Armstrong, H. A., Williams, M., Zalasiewicz, J. A., and Sabbe, K. (2009). Ground-truthing Late Ordovician climate models using the paleobiogeography of graptolites. *Paleoceanography* **24**, PA4202.

—— —— —— —— —— Paris, F., Nõlvak, J., Challands, T. J., Verniers, J., and Servais, T. (2010). Polar front shift and atmospheric CO2 during the glacial maximum of the Early Palaeozoic Icehouse. *PNAS* **107**, 14983–6.

van der Hammen, T. (1974). The Pleistocene changes of vegetation and climate in tropical South America. *Journal of Biogeography* **1**, 3–26.

Van Kranendonk, M. J. (2011). *Science* **333**, 413–14.

Vaughan, A. P. M. (2007). Climate and geology—a Phanenozoic perspective, in M. Williams, A. Haywood, F. J. Gregory, and D. Schmidt (eds) Deep-time perspectives on climate change: Marrying the signal from Computer Models and Biological Proxies. (London: The Geological Society for the Micropalaeontological Society), 5–59.

Wacey, D. (2010). Stromatolites in the approximately 3400 Ma Strelley Pool Formation, Western Australia: examining biogenicity from the macro- to the nano-scale. *Astrobiology* **10**, 381–95.

Wade, B. S. and 8 others. (2012). Multiproxy record of abrupt sea-surface cooling across the Eocene-Oligocene boundary in the Gulf of Mexico. *Geology* **40**, 159–162.

Walker, M. J. C. (2004). A Late glacial pollen record from Hallsenna Moor, near Seascale, northern England, with evidence for arid conditions during Loch Lomond (Younger Dryas) Stadial and early Holocene. *Proceedings of the Yorkshire Geological Society* **55**, 33–42.

——and 17 others. (2008). The Global Stratotype Section and Point (GSSP) for the base of the Holocene Series/Epoch Episodes **31**, 264–7.

Wallace, D. J., and Anderson, J. B. (2010). Evidence of similar probability of intense hurricane strikes for the Gulf of Mexico over the late Holocene. *Geology* **38**, 511–14.

Wang, X., Edwards, R. L., Cheng, H., and Shen, C-C. (2004). Wet periods in northeastern Brazil over the past 210 kyr linked to distant climate anomalies. *Nature* **432**, 740–3.

Wang, Y., Cheng, H., Edwards, R. L., Kong, X., Shao, X., Chen, S., Wu, J., Jiang, X., Wang, X., and An, Z. (2008). Millenial and orbital-scale changes in the East Asian monsoon over the past 224,000 years. *Nature* **451**, 1090–3.

Watanabe, T., and 11 others. (2011). Permanent El Niño during the Pliocene warm period not supported by coral evidence. *Nature* **471**, 209–11.

Wilde, S. E., Valley, J. W., Peck, W. H., and Graham, C. M. (2001). Evidence from detrital zircons for the existence of continental crust and oceans on the Earth 4.4 Gyr ago. *Nature* **409**, 175–8.

Willett, S. D. (2010). Late Neogene erosion of the Alps: A climate driver? *Annual Review of Earth and Planetary Sciences* **38**, 411–37.

Williams, J. H., De Benedictis, A., Ghanadan, R., Mahore, A., Moore, J., Morrow, W. R. III, Price, S., and Torn, M. S. (2012). The technology path to drop greenhouse gas emissions cuts by 2050: The pivotal role of electricity. *Science* **335**, 53–9.

Williams, M., Siveter, D. J., Ashworth, A. C., Wilby, P. R., Horne, D. J., Lewis, A. R., and Marchant, D. R. (2008). Exceptionally preserved lacustrine ostracods from the middle miocene of Antarctica: implications for high-latitude palaeoenvironment at 77° south. *Proceedings of the Royal Society*, **B275**, 2449–54.

Williams, M., Nelson, A. E., Smellie, J. L., Leng, M. J., Johnson, A. L. A., Jarram, D. R., Haywood, A. M., Peck, V. L., Zalasiewicz, J. A., Bennett, C., and Schöne, B. R. (2010). Sea ice extent and seasonality for the Early Pliocene northern Weddell Sea determined from fossil *Austrochlamys* bivalves. *Palaeogeography, Palaeoclimatology, Palaeoecology* **292**, 306–18.

Wilman, R. A., Jr., Hickey, L. J., Dickinson, M. B., Berner, R. A., Robinson, J. M., Dietrich, M., Essenhigh, R. H., and Wildman, C. B. (2004). Burning of forest materials under late Palaeozoic high atmospheric oxygen levels. *Geology* **32**, 457–60.

Wilson, R., Cook, E., D'Arrigo, R., Riedwyl, N., Evans, M. N., Tudhope, S., and Allan, R. (2010). Reconstructing ENSO: the influence of method, proxy data, climate forcing and teleconnections. *Journal of Quaternary Science* **25**, 62–78.

Witze, A. (2012). Climate change confirmed... again. *Nature Geoscience* **5**, 4.

Young, G. M., von Brunn, V., Gold, D. J. C., and Minter, W. E. L. (1998). Earth's oldest reported glaciation: Physical and chemical from the Archaean Mozaan Group. *Journal of Geology* **106**, 523–38.

Zachos, J. C., Opdyke, B. N., Quinn, T. M., Jones, C. E., and Halliday, A. M. (1999). Early Cenozoic glaciation, Antarctic weathering, and seawater 87Sr/86Sr: is there a link? *Chemical Geology* **161**, 165–80.

—— Pagani, M., Sloan, L., Thomas, E., and Billups, K. (2001). Trends, rhythms, and aberrations in global climate 65 Ma to present. *Science* **292**, 686–93.

—— et al. (2005). Rapid acidification of the ocean during the Paleocene-Eocene thermal maximum. *Science* **308**, 1611–15.

—— Dickens, G. R., and Zeebe, R. E. (2008). An early Cenozoic perspective on greenhouse warming and carbon-cycle dynamics. *Nature* **451**, 279–83.

Zalasiewicz, J. A., Taylor, L., Rushton, A. W. A., Loydell, D. K., Rickards, R. B., and Williams, M. (2009). Graptolites in British stratigraphy. *Geological Magazine* **146**, 785–850.

—— Williams, M., Steffen, W., and Cautzen, P. (2010). The new world of the Authropocene. *Environmental Science and Technology* **44**, 2228–31.

Zanazzi, A., Kohn, M., MacFadden, B., and Terry, D. O., Jr. (2007). Large temperature drop across the Eocene–Oligocene transition in central North America. *Nature* **445**, 639–42.

Zimov, S. A., Schuur, E. A. G., and Chapin, F. C. III (2006). Permafrost and the global carbon budget. *Science* **312**, 1612–13.

Index